ALLEN COUNTY PUBLIC LIBRARY
3 1833 00269 5184

S0-BWM-913

The Solar Jobs Book

The Solar Jobs Book

How To Take Part in the New
Movement toward Energy Self-Sufficiency

Katharine Ericson

Brick House Publishing Company
Andover, Massachusetts

Published by Brick House Publishing Co., Inc.
3 Main Street
Andover, Massachusetts 01810

Production credits
Editor: Jack Howell
Book design: Kenneth Wilson
Cover design: Susan R. Z. Slovinsky
Typesetting: Neil W. Kelley
Production supervision: Dixie Clark

Printed in the United States of America

Library of Congress Cataloging in Publication Data
Ericson, Katharine, 1922-
The solar jobs book.

Bibliography: p.
Includes index.
1. Solar energy industries—United States—vocational guidance. I. Title.
HD9681.U62E74 331.7′62147′0973 80-17886
ISBN 0-931790-12-3 (pbk.)

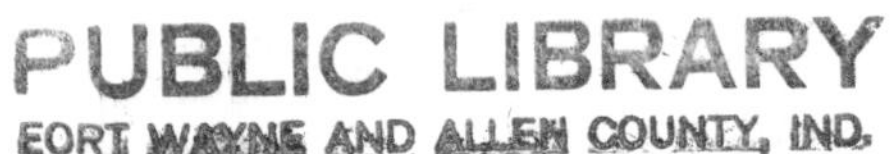

"Energy is destined to play an increasingly visible role in the shelters of all people everywhere. While some are utilizing the latest advances in photovoltaic technology, others will be reasserting the ancient wisdom of planting shade trees and windbreaks, of harnessing prevailing winds for ventilation, and of relying on thick ceilings and walls to even out daily extremes in temperature. The most successful in all cultures will be those who realize that we have reached the end of an era, and who design shelters to work with nature instead of defying it."

Denis Hayes
Rays of Hope: The Transition to a Post-Petroleum World[1]

Contents

Introduction

Although the sun has been in its place in the heavens for some 4,000 million years—and promises to be there for at least 5,000 million more—relatively little has been learned about the star that is the earth's orbital governor and its life support. There is polite, but spirited, scientific debate about the sun's basic dimensions; it is "about" 93 million miles from earth, and "about" 865,000 miles in diameter. Equally prominent individuals in the fields of physics, astronomy, and so forth, will tell you with equally convincing conviction that the sun is growing, shrinking, moving away, moving closer.

The debates are typical of recent times in this nation and the world. During the past century we have witnessed the birth and maturity of the age of science and technology. Physics has prevailed over metaphysics; the ancient and respected practice of sun worship has been derided; and humankind's relationship with the solar presence has been reduced to the often unsuccessful attempts of various experts to obtain solar data, as opposed to the process of developing solar insights.

The times they are a-changing, however. As the inevitable diminution of fossil fuels and nonrenewable resources has come to pass with accelerating reality, much of the populace has begun to develop a consciousness of limits. Recognizing the futility of planning even grander petro-futures, and understanding that some day even coal and natural gas will be scarce, a good many young people (and some older ones) have begun to think clearly about renewable resources—of which the sun is the primary one.

Solar-oriented thinkers have left most of the technologists far behind. While arguments rumbled over how much crude oil might be buried beneath the ocean floor, a few people began, on their own, with insights rather than data, to construct solar living spaces, design solar heaters and coolers, and develop the base for the solar metaphysics that must be in place before a Solar Age can be established. A process had begun.

This book is a key part of that process. It has been written and published by individuals who have thought creatively about the sun instead of debating endlessly its precise dimensions. Katharine Ericson and Brick House have taken a major step toward helping all of us re-establish a significant relationship with our star.

Solar jobs are not listed on the bulletin boards of most employment agencies, nor do they loom large on the statistical matrix at the Department of Labor. Nevertheless, they are among the jobs of the future—everyone's future. That is an obvious, albeit metaphysical, insight which all too few of us have recognized.

Thank goodness Kay Ericson did. What we have here is a beginning, a start at helping would-be solar engineers, architects, attorneys, carpenters, financiers, teachers, researchers, writers, executives, and, yes, even solar bureaucrats. This book is a guide to their solar careers; it is a first; it is as much an insight as it is an index.

And, it is important. There must be tens of thousands of young people in this nation right now, in cities and towns and on campuses, who are quite aware that their futures will include significant solar relationships. They are interested; they want to work; they comprehend the necessity. Now they have a guide to help them begin.

In the late sixties, Bill Shurcliff, a friend, began compiling a directory of solar-heated buildings. His was an individual effort, home-typed, mimeographed and mailed to a few interested persons. By 1978, that directory had become too much of a job for one person.

The beginnings of the Solar Age have consistently been sparked by individuals like Bill Shurcliff, not by organizations. *The Solar Jobs Book* is the latest and one of the brightest of those sparks, and I am quite certain that this volume is also only a beginning. Hopefully, a dozen editions from now, the task of coordinating and listing the solar jobs available will have grown too large for one writer—even Kay Ericson.

For now, however, she has made an excellent start.

John N. Cole
Brunswick, Maine

Part I

Working Where It Counts

The Detached Mood of the Current Work Force

When Studs Terkel's *Working*[1] was published in 1974, most of its readers were relieved to find that they were not the only ones who found a lack of meaning in their jobs. In discussing their attitudes toward their work, people from all walks of life told Terkel "the system stinks"; "our jobs are too small for our spirits." He was impressed with the number of people who, despite their disenchantment with their jobs, managed to retain a strong sense of personal worth and a hope of finding ways to work more creatively.

Some people are finding ways in opportunities that have been created by impending changes in the very foundation of our society—in our supply and use of energy. These are the people who have grasped the potential of the immense energy emanating from the sun. As the world becomes increasingly concerned about dirty, dangerous, and diminishing fuel supplies, they are challenged by the development and wise use of solar energy in all its forms, including the wind, water, and the other natural systems that it feeds. They find meaning in solar jobs because of the compelling need for clean, renewable sources of energy. They want to use this kind of energy to help restore a measure of self-sufficiency to people who have become increasingly dependent on the large, remote systems of a highly industrialized society.

So far, very few resources exist to help people understand and find employment in the development of this Solar Age. Its beginnings reach into so many aspects of our culture that we cannot deal with all aspects of solar jobs at one time. Some people are working on using the wind, the ocean, or municipal wastes to generate electricity; others on making alcohol from grain to serve as automobile fuel. Electric power and transportation are only two of the several energy-based systems that we have come to need. One human need is even more basic—the need for shelter. For this reason, *The Solar Jobs Book* emphasizes the use of solar energy in buildings. It begins to answer some very practical questions that are heard repeatedly: Where are the solar jobs? How can I get into them?

Unfortunately, the "efficient" systems that were developed in the times of cheap energy have led to an intense specialization of the work force. The consequences are powerfully presented by Wendell Berry, a Kentucky farmer and writer:

> A system of specialization requires the abdication to specialists of various competencies and responsibilities that were once personal and universal. Thus, the average—one is tempted to say, the ideal—American citizen now consigns the problem of food production to agriculturalists and 'agribusinessmen', the problems of health to doctors and sanitation experts, the problems of education to school teachers and educators, the problems of conservation to conservationists, and so on. This supposedly fortunate citizen is therefore left with only two concerns: making money and entertaining himself. He earns money, typically, as a specialist, working an eight-hour day at a job for the quality or consequences of which somebody else—or perhaps, more typically, nobody else—will be responsible. And not surprisingly, since he can do so little else for himself, he is even unable to entertain himself, for there exists an enormous industry of exorbitantly expensive specialists whose purpose it is to entertain him.

Berry continues,

> . . . The fact is, however, that this is probably the most unhappy average citizen in the history of the world. He has not the power to provide himself with anything but money, and his money is inflating like a balloon and drifting away, subject to historical circumstances and the power of other people. From morning to night he does not touch anything that he has produced himself, in which he can take pride.[2]

Many of the people who are working on the development of solar energy are moving away from the current worship of specialization. They prefer to cross disciplines, to move back and forth between physical and mental labor, and to blend their work and personal lives. They view as humane and psychologically healthy people using more of their own energy on behalf of their own maintenance. Growing a small amount of food in a backyard garden and bicycling to work, for instance, not only can conserve energy but can replace joining a health spa to keep physically active. These people believe that even small efforts can increase an individual's sense of wholeness and independence. Developers of solar energy also hope to affect neighborhoods, communities, and regional areas by enabling people to use the renewable energy resources available to them right where they are.

We have only recently recognized that many of our systems are wasteful, impersonal, and too big—largely as a result of cheap energy.

When energy cost little and was considered limitless, there seemed to be no end to how efficient we could become, how big we could grow. With natural gas and oil at our command, clearly it was relatively efficient to build a new structure with windows that could not be opened so as to afford continuous air conditioning in the summer and heating in the winter. Rather than sell his crops to his neighbors, it was more profitable for a farmer in northern New England to ship his crops to Boston or New York in trucks, where they were packaged, and then sold to a distributor, who shipped them back to be sold in the store in the farmer's home town. The examples of unnecessary use of energy are legion, from "throw-away" packaging of consumer products to "throw-away" heat issuing from steel mill smoke stacks high in the air.

It was easy for people to grow accustomed to the "throw-away" style. And, by the mid-1970s, the United States, with 6 percent of the world's population, was using 40 percent of the world's energy sources.[3] Some people were dimly aware that they were getting farther and farther away from understanding, much less controlling, the fundamental sources of their food, shelter, and employment. But why worry?

The Beginnings of Change

On the day in 1973 when the Arabs stopped shipping their oil to the United States, the "new energy era" started in this country. Although the realities of our energy supply and our energy consumption had been there all along, the Arab Oil Embargo had a startling effect on the public consciousness. As if awakening from a long sleep, people saw for the first time that our entire society was dependent on fuel from foreign sources and that we were vulnerable. They became painfully aware that they could hardly even trace the sources of supply of food and heat, much less exert any control over political or economic upsets that might stop food from being trucked across the continent or heating fuel from being delivered. A new term moved into the vernacular—"fossil fuels." Coal, oil, and natural gas, people now understood, took millions of years to form. And once these fossil fuels were used up, it would take millions of years for more to form.

When Arab oil started flowing again, at much higher prices, some people managed to ignore the situation, to go on with life as usual, to believe that nothing was really going to change much after all. Others, realizing that consumers must be part of a problem involving consumption, did what they could to lower energy use, from lowering thermostats to buying smaller cars. They attempted to remain informed and open to any adjustments and changes that the energy situation might require. By now, it is clear to large numbers of people that we have left the age of cheap energy and have entered a transition phase, moving toward new sources of energy and ways of using them that eventually must call for new ways of living and working.

A change of this magnitude is not made in a year or in a decade. We will continue to use traditional sources of energy, but in ever decreasing amounts. Fortunately, the emerging changes in our overconsuming and energy-intensive society are not necessarily negative or depriving ones. It is as if the natural order has, by the finite nature of the fossil fuel supply, put a check on the "throw-away" trend with its high monetary, and sometimes high psychological cost of what we call modern "comfort" and "convenience."

During this transition time, effort continues to go into the development of the remaining sources of fossil fuels. In fact, long before the Arab Oil Embargo, a number of people had been working on assessing the extent of our remaining fossil fuels, as well as on the development of alternative energy sources of all kinds. The Continental Oil Company has more than doubled its hiring of college graduates since 1973, a fact that, according to the personnel manager, is "a direct result of the massive dollar investments we are making to find and produce new energy sources." There will be increased offshore drilling; exploitation of secondary and tertiary layers of oil, of shale oil, and the development of more projects like the Alaskan Pipeline. Coal will be revived. There may be some further development of nuclear power plants, although their exorbitant construction costs and questionable safety standards may cripple their development. Many people will find satisfaction and interest in continuing to develop such sources as long as possible.

Once you begin to talk with people who are in the forefront of shaping the transition, you begin to find the excitement. The national Solar Domestic Policy Review, submitted to President Carter last year, estimated that, with the proper government programs, renewable energy resources could contribute more than 25 percent of the nation's energy by the year 2000. Even if such resources were to account for only 10 percent of the energy market in the next two decades (a figure sometimes predicted by very conservative estimators), that, at today's prices, would save billions of dollars a year in oil imports.

People working in the solar energy field foresee that eventually the country must be based more and more on renewable resources and careful consumption. They are in on the beginning of the transition, helping to develop a variety of energy-saving, safe technologies and related new social and economic patterns that can lead toward a high quality of life.

Change to What?

Who could have predicted, even a few years ago, that in 1978 our nation would participate in a Sun Day? On that day (May 3, 1978) thousands of local events, financed by individuals, foundations, unions, and the government, and fully covered by the media, demonstrated wide popular support for the development and use of solar energy. The day was organized by Denis Hayes[4] and other people, who, eight years previously, had established Earth Day to encourage the protection of our environment from pollution. While Earth Day alerted us to a set of environmental problems that had been ignored for decades, Sun Day emphasized positive, attainable solutions to energy problems that only recently had become apparent. Sun Day and other happenings helped us to acknowledge the gravity of the energy problem and to move on a solution, which will be found, in large part, through applying the twin concepts of *conservation of energy* and *use of renewable sources of energy* to help augment fossil fuels.

People do not yet apply these concepts regularly in their daily lives, but there is an uneasy feeling that the present energy life style cannot continue forever. Today, some of us may fill our gas tanks and drive as far as we like, but we know that many people in other localities and-or circumstances cannot—what with gasoline shortages and prices soaring over a dollar a gallon. Today we may be able to get together the money to heat a house with leaky windows and inadequate insulation, but we know that the supply and price of fuel ultimately will force us to do caulking, weatherstripping, and so on, to conserve the units of fuel we use; and, on a larger scale, to find sources that, unlike fossil fuels, will not take millions of years to form and will be continually renewable.

Many people are unaware of how much progress already has been made. The methods and institutions that solar workers have begun to work with are forming a nucleus for the growth of permanent solutions during the next decade.

Solar energy, long considered only a dream of some environmentalists, is quietly becoming an industry in the province of corporations, labor unions, small business people, workers, and others concerned with economic development. Evidence of its existence is found in all parts of the country. In areas where there are large numbers of people with enthusiasm and expertise, networks of enterprises and support systems have grown up. California's solar industry, consisting of more than 300 firms, is the largest in the United States today. Many of the large oil companies, as well as other major companies, are beginning to invest their time and assets in the solar field, either to create systems to provide energy from solar sources or to develop applications for the use of that energy. In 1978 the Sears, Roebuck catalog heralded the advent of mass marketing for solar technologies by carrying for the first time solar equipment for hot water and space heating.

The federal government got off to a slow start in this field, but now is advancing. In 1976 a meeting to organize a Congressional Solar Coalition attracted only six congressmen; by 1978 the coalition had become a group of more than 100 members of the House and Senate, with a broad legislative package. When President Carter included only $370 million for solar research in the 1979 budget, Congress increased the amount to $529 million. The President's 1981 budget request asks for $1.1 billion for solar research.

Government money is used to finance and encourage not only technical research, but also commercial development and the education of the public. Grants have been awarded for installation of renewable energy systems in private homes, schools, hospitals, and housing developments. Several states, aided by federal funds, have enacted their own tax credits for individuals who install such systems in their homes and businesses.

In a supportive move, the federal government announced one week before Sun Day 1978 that it would install a solar hot water system in the White House. Subsequently, the House of Representatives authorized the installation of a solar heating system in the Rayburn House Office Building. Long before this, other federal, state and county agencies had initiated programs to prevent heat loss in their buildings and to install solar equipment.

The solar industry and the number of people working in it is growing rapidly in response to three significant trends: 1) use of government incentives to encourage the use of energy-efficient methods and solar equipment in homes, work places, and on farms; 2) increases in the costs of fossil fuels; and 3) decreases in the costs of solar energy systems.

Many recent publications analyze our energy problems and predict our energy future. Three of these express clearly the theoretical assumptions on which this book is based. Amory Lovins' *Soft Energy Paths*[5] presents the specific technologies and social planning that will make possible, by the beginning of the next century, energy based mainly on renewable resources. E. F. Schumacher's *Small Is Beautiful*[6] has a subtitle that points up its contribution: *Economics as if People Mattered*. Wendell Berry's *The Unsettling of America*[7] gives a closely reasoned analysis of the psychological and social need to reverse our present trend toward big, centralized, and remote systems. In combination, these authors chart technological, economic, and social goals for the future and show how they can be reached. Robert Stobaugh and Daniel Yergin in *Energy Future* and the Union of Concerned Scientists' *Energy Study* document similar material.[8]

The Change Makers

It is clear that the development of energy conservation measures and solar technologies will create jobs, rather than reduce employment, as has been predicted by the advocates of large, centralized electric power plants. A Federal Energy Administration analysis[9] estimates the effect on employment if only 37,372 private homes were provided with simple energy saving improvements such as insulation, automatic thermostats, and efficient furnaces. The analysis shows that, over a seven-year period, 488,000 jobs would be created: 122,000 in manufacturing and 366,000 in local installation. The report stresses that employment associated with energy conservation requires low to moderate skill, and tends to be available near urban areas, where the most acute unemployment problems occur.

According to Edward J. Carlough, President of the Sheet Metal Workers' International Association:

> At a relatively low cost, we could make a huge dent in the nation's energy consumption by modifying those systems to use solar energy where it is feasible. The national value of a concerted effort in solar energy and energy conservation would be measured in millions of barrels of oil—resulting in more favorable balances of payments, less inflation, more jobs and energy independence. . . . even figured conservatively, energy-saving modification work and an expanded use of solar energy could put all unemployed sheet metal workers back to work.[10]

A comprehensive government study, prepared for the Joint Economic Committee of the U.S. Congress and released in April, 1979, estimates that 3 million jobs could result if the government planned a massive shift from oil and coal to solar by 1990. This is one of a number of careful and detailed studies done by different government and private researchers on the effects of solar energy on employment.[11]

The U.S. Department of Labor has listed some of the skilled workers needed to build and maintain solar units: carpenters, cement masons, bricklayers, painters, welders, iron and insulation workers, surveyors, electricians, plumbers, sheet metal workers; air conditioning, heating, and refrigeration technicians; glaziers; and crane operators. People are also needed in occupations such as solar engineering, architecture, law, banking, real estate and appraisal, sales, zoning, assessment, and consumer protection.[12]

People in the skilled trades already are engaged in solar industries, and their numbers are growing. Many have not experienced much other change in their lives and may not yet be fully aware that their work is part of a long-term changing energy pattern. But the trades and the other occupations mentioned above also are beginning to attract new workers whose primary purpose is to become involved with the development of solar energy.

The paths to satisfying vocations in solar energy are many and can be highly individual. In terms of personal economic security, some paths are more risky than others. Many challenge a person to use self-knowledge and imagination, both in finding a way to start and in continuing to create his or her own place. A passive, detached, and sometimes almost despairing attitude toward finding a job—and toward working itself—is not characteristic of people who choose to work on renewable energy forms. They did not fall into their work by accident. They sought out work that they considered meaningful and often they began by doing tasks they never dreamed they could or would do.

Some of those committed to working toward a society that conserves its resources and uses them wisely are subject to nostalgia, an idealization of the small community of the last century. If they think seriously of actually returning to such a society, they realize that they are indeed "modern" people and that such a return would be not only impossible, but also not altogether desirable. Some of the vocations, occupations, and life styles described in this book may seem to resemble aspects of older agricultural periods in our history; in fact, they are only approximations.

Many of the people in these new energy-related occupations share certain goals: getting energy from a small regional power plant or (the ultimate in being close to the source of supply) from a solar panel or solar cell directly on the roof of a house; growing and-or buying food from sources close to home; dispensing with wasteful consumption; and working in small units that are more directly connected with their own communities.

Although people of all ages are entering the solar jobs field, workers at the early stages of their careers have more to gain, for they are part of the "baby boom." The large numbers of children born after World War II, who swamped the school systems, are now starting to reach the offices and the factories. In 1975 there were 39 million workers in the 25 to 44-year-old age bracket (considered the prime years for job development); in 1990 there will be 60.5 million workers in that age bracket.[13]

This bunching of the work force will have far-reaching consequences, not the least of which will be fierce competition for promotions and good jobs in the next decades. An increasing number of people will experience disappointment and frustration if they decide to try moving up the organization ladder. Rather than competing for scarce places, many choosing the solar energy field are striking out in new directions where ladders haven't even been built yet.

Part II

A New Breed of Workers

New Values, Work, and Life Styles

The New England change of seasons means more to people working in solar energy than just a change in the weather. For those who have established themselves in the field of energy-efficient building, it not only means knowing how buildings interact with sun, wind, and temperature, but it also means mixing the various aspects of their work differently according to the time of year.

When, for instance Jeremy Coleman, of Marlboro, Vermont, goes to work in the summer, he takes his carpentry tools with him. On a typical morning, he does construction work at the site of the new house that he and his partner are building. Careful and skilled craftsmen, the partners make sure their construction is tight so that no air will leak out. They will install the windows, purchased from a large and well-known maker of double-paned thermal glass, with the same attention they give to the heavy oak front door.

These builders know not only about air leaks, but also about what insulation gives maximum protection in walls and attic, how the building should be placed on the property in order to be benefited by the sun in the winter, and what kind of overhang will protect it from the sun in the summer. All these bits and pieces of information are part of their store of knowledge about passive solar heating—taking advantage of all the natural factors and traditional materials to save energy in a building (especially important in the northern climates).

At noon Coleman leaves to check on another house under construction for which he has designed the installation of solar roof panels. Here, he eats lunch with one of the workers, a young man who is gaining experience as Coleman did himself a few years earlier. In this example of passive solar heating, as in others, the heat from collectors is circulated through the house without moving parts and without provision for storing the heat. A simple closing or opening of ducts keeps heated air in or moves it out. This is a type of building that would not have been built five years earlier. Even now, only a handful of builders and architects in southern parts of Vermont and New Hampshire are able to design or supervise such construction.

The two men take a longer-than-usual lunch hour. Both college graduates, they talk about a best-selling novel by a local writer, and about the Marlboro Music Festival, which is about to begin. Then they start talking business—about the best sources of supply for certain materials, the costs of "active" solar systems, the tricks of installing a solar collector properly.

After lunch, Coleman proceeds to the home of a couple who are considering the construction of a greenhouse on the South side of their house. He advises them on that and also on adding insulation in the attic so as to save heat. Most home owners are living in houses built during the time of cheap energy, when little attention was given to saving fuel. Now many owners are asking where they can find someone who will advise them on making energy-efficient changes to existing buildings, a process called "retrofitting" in the trade. Advice includes estimating the actual dollar savings that would be effected by specific changes and comparing savings with the costs of installing the improvements. With this information, home owners can figure out how long it will take to pay back their investment with the money they save in fuel.

After a short time back at the site of the new house, Coleman heads home. He is physically tired and ready to clean up and to relax. He picks a few vegetables from his garden for dinner. He looks at his mail, which includes specifications for new solar greenhouse components—specifications he is anxious to read, since he has started to specialize in such construction—and also a new book on solar construction that he plans to recommend to others.

On a winter morning, Coleman carries in wood and stokes the stove to heat his house while he is gone. If the sun is out, he pulls back the curtains and raises the shades, (drawn and lowered during the night to help retain the heat of the day). Then he gathers together folders containing tables, specifications, and other data, leaving his tools at home. He goes to the offices of the Brattleboro Design Group, where he works on designs for passive solar houses, not on actual construction.

Last winter, Coleman worked at Total Environmental Action (TEA), Inc., a New Hampshire solar design and consulting firm, where, along with others, he did the research and design for a low-cost solar panel, a project funded partially by the federal government. In this type of setting, his work is, of course, more theoretical than his work on the construction site, but his experience in building and in solar installation enables him to know how engineering principles will work out in the field.

During the school year, Coleman left TEA early so that he could prepare for teaching an evening seminar on using solar energy in the home, a course offered by a local college. He also teaches a course on solar greenhouses at TEA on Saturdays. Coleman's teaching demonstrates a sensitivity to students, an ability to present material clearly, and an integration of theory and practical application. For Coleman and his colleagues, in research, in design, and in construction, reading and discussing the latest information on the economics and technology of energy-efficient building are part of a way of life. They are continually learning new things. Theory and practical application are both important and a grasp of both is characteristic of the people working in the solar energy field.

People attending Coleman's courses include college students, home owners, architects, builders, fuel dealers, bank loan officers, and real estate agents, many of whom need to know about the new trends in building and solar heating.

The development of Coleman's career is an example of someone's following up on a subject that is of interest and importance to him or her. As an undergraduate at Marlboro College, Coleman became interested in the use of solar energy and he designed a course of study to introduce himself to the field. He also found an internship, one of the best methods *of getting connected*, as well as getting experience. Coleman's internship working with solar designers at TEA led to his part-time employment there. He did not know exactly where his interest would lead: his priority was the subject of solar energy rather than a well-defined and secure career path.

Although Coleman's particular mix of activities may not be duplicated by anyone else, his present work, education, and life style have much in common with those of the growing numbers of people who are entering the solar building and heating field.

In sum, Coleman's work involves a combination of occupations, including carpentry, designing, teaching, consulting, research, and business. It is both physical and mental. It has been fostered by Coleman's seeking out diverse people and activities in his own particular geographic region, and it enables him to move around to various settings, from the construction site to the classroom or research lab. Coleman directs his own education, continuing to study and to learn from both traditional and nontraditional sources; people with experience and information are an important part of his continuing education. He uses his formal education, both technical and liberal, in his work and personal life.

Coleman's life style is consistent with the principles of the field he has chosen, remaining, as it does, close to the sources of supply for his food and other needs. He chooses simplicity and informality—for instance, taking his lunch to work and rarely changing clothes as he goes from one activity to another. Time off and time on are flexible. Instead of putting in regular hours each week at times decided by an employer, Coleman himself assumes the responsibility for getting his work done. His income is moderate, and since it is derived from more than one source, he cannot predict its exact amount in any given year.

Many of Coleman's contemporaries have found jobs in large, established industries that are now turning segments of their businesses toward developing better solar collectors, manufacturing solar hot water heaters, or selling products such as insulation and storm windows. These people, who are described in some detail in Part III, usually work on a straight salary and live in or near cities of at least medium size. The new kinds of workers, on the other hand, are distinguished from those going a more traditional route by their independence from large organizations and their willingness to take risks. Often they have started by choosing a smaller community as their base.

Within a smaller community, or more accurately, a constellation of small communities in a given geographic area, the individual solar worker ties into a network of small enterprises and like-minded people which replaces the support system usually provided by the large employer. An independent builder, running into problems of putting solar panels on an old house without destroying the aesthetic integrity of the building, is unable to turn to a corporate design department. Instead, he knows that an older architect who has become a specialist in solar design has recently moved into the area, working out of his remote country home. He knows this because he gets his thermal glass windows from a small supplier specializing in high quality materials for solar homes. Located in a renovated old church building, this solar materials store has become the gathering place for the new energy crowd, and the unofficial dissemination center for information on who is doing what. Since most of the people frequenting the store work on their own time schedules, there is time to stop and talk shop, to get to know each other, to read the bulletin board and the professional magazines on the table. There will probably even be time to arrange an informal gathering for Saturday night.

Networks such as this one now exist, at least on a small scale, in almost every part of the United States. Sometimes they are not highly visible to the rest of the population; in certain areas of the country they have become not only visible but quite well developed. Solar building activity is particularly strong in Arizona, California, Colorado, New Mexico, and the New England states. In these places, although the primary contacts are still the local ones, a network stretches out through an entire state or region. People move easily from one region to another, once they have gained experience.

In order to gain experience, people who start only with a knowledge of solar energy must be willing to learn specific occupations where that knowledge can be applied. On the other hand, workers who already have a skill or profession can find ways to apply it to furthering the new energy forms. A carpenter builds solar collectors. A writer edits a newsletter for a solar industries organization. A state official administers tax incentives for homeowners who add solar devices. A film maker makes training films for installers of hot water heaters. A consultant makes an evaluation of heat losses in a factory. An architect designs an energy-efficient house.

Although the use of solar energy is a technical or scientific subject, an understanding of its general principles requires only a common-sense intelligence. Basically, its principles are a matter of how to convey heat from the sun into a building during cold weather and to keep as much of it there as long as possible; and, in hot weather, of course, the reverse: to keep the sun's heat outside of a building. Some techniques, such as the placing of a house properly on its site or the very important caulking of tiny gaps around windows and doors, can be grasped easily by the most "unhandy" handyperson. Detailed technical knowledge of this sort is all that is needed for many jobs in the field, and it is available to anyone willing to read books or to attend short-term seminars at a community college. In fact, the proliferation of publications and courses on the subject suggests that information on using solar energy may be in itself a growth industry.

Imaginative people who are already working in businesses related to building or heating sometimes turn out to be the most enthusiastic and effective self-trained solar promotors. A few years ago the 40-year-old service manager for a small oil distributing company began to read up on forthcoming changes in oil supply. He realized that his company, rather than simply delivering oil, could move toward offering a total "energy management" program. First, he added hot water heaters to the company's line of furnaces and heating equipment; then, storm doors and windows, and insulation; and, more recently, solar hot water heaters. He took seminars on solar energy, and traveled, at least once during a vacation, to different areas to observe different methods of solar heating.

This service manager not only trained himself, but subsequently he saw to it that the company managers paid for several of their young workers to take basic solar energy courses and installers' workshops. The company now can provide workers to evaluate for a customer ways in which heat can be supplied and conserved most efficiently in that customer's home or business. Now, instead of simply bracing themselves for rising prices and short supplies of oil, the managers are actively developing a business based on the realities of the next few decades.

People in the building industry may have the greatest opportunity to provide shelter based on renewable resources and conservation of energy. Sometimes they are also the most resistant to change, being, as many of us are, comfortable with familiar methods. They may quote a news item to the effect that solar systems are exorbitantly priced and maintain that they at least will wait until the systems are cheap and well established. Usually they are referring to elaborate "active" solar heating systems that require large areas for storage of heat, motors and fans to move the heat, and finely tuned thermostats or control mechanisms to regulate it. Indeed, these active systems are not yet economical, but many steps can be taken short of a completely active system. Like many oil distributors, the plumbers, carpenters, contractors, and suppliers of building materials who are informing themselves about and starting to use solar principles are getting a head start.

In the smaller communities, the solar energy work force is a mixture of natives and newcomers. Newcomers from nonrural areas seem to have little trouble with being accepted if they are willing to work and to learn. All of these workers have the common bond of pioneering in a new and rapidly changing field, where they need one another's input.

Since 1970, for the first time in the twentieth century and probably in the history of the country, rural areas have been growing faster than metropolitan ones, according to the Census Bureau. A report by a government demographer states:

> The vast rural-to-urban migration of people that was the common pattern of the U.S. population movement in the decades after World War II has been halted and, on balance, even reversed. In the eyes of many Americans, the appeal of major urban areas has diminished and the attractiveness of rural and small town communities has increased.[1]

Due to this migration (a gain of 4.2 million people in rural counties between 1970 and 1973 alone), many rural towns no longer seem so "rural." Small industries, retirement and vacation homes, tourism, and shopping malls have tended to bring some of the urban and suburban ways of life to the country. The quality of life is still different, however. A pace more closely tied to nature's seasons replaces the hustle and anonymity often imposed by the city, and individual resourcefulness and self-reliance are in demand.

The "trade-off" does not appeal to everyone. It usually involves lower salaries and less access to services and material goods. Although those choosing to settle in small communities do not always find these compromises easy, they prefer living in a small-scale environment, one that the human mind can encompass. Most of the "migrants" working in solar building have achieved a self-sufficiency they formerly lacked, but, contrary to popular opinion, only a very small number of them are counter-culture young people attempting to homestead.

Some of the new energy people are older professionals, such as architects, engineers, or business people who have built up some measure of financial security. The younger workers, however, the ones who are just starting out, are foregoing financial security while they learn a new skill or business. A young engineer may find himself building a house; a history major may find herself bookkeeping for a solar building supply company. No doubt he and she could make more money elsewhere, but they are gaining experience in the solar energy field and plan to expand their experience in the future.

The younger people are finding the opportunity to learn primary productive skills a welcome change from years of academia. The college educated are discovering the dignity of the trades and of intelligent physical labor. They are being rewarded primarily not by money but rather by enjoyment of the learning process and the importance of their commitment.

New Settings for Design, Research, and Education

As many people find their places within existing communities, others, a small number, are becoming associated with new kinds of communities that are committed to renewable energy forms. Almost immediately after the Arab Oil Embargo, groups formed to establish places where work could be done on practical, ecologically sound alternatives to a dependence on fossil fuels. Unlike many other research and development projects started in recent years, these groups did not become subgroups within universities, industries, or government agencies, but were set up independently, reflecting the fact that the founders were working with concepts considered unimportant in most traditional settings. They were making a conscious attempt to build models for the more decentralized, self-reliant households and neighborhoods that they believe will evolve along with the wise use of energy.

One of the first places in the United States to attempt to redesign our basic life support systems so as to utilize such renewable resources as sun, soil, wind, and biological growth and decay was the nonprofit Farallones Institute in Northern California. The institute is concerned with making these systems yield a high quality of life and with making them possible in both urban and rural areas. An important part of the institute's program is the establishment of residential centers, where people actually experience the environment created by the ecological processes they advocate.

The Farallones Urban Center is located in a mixed industrial-residential neighborhood in Berkeley, where the Integral Urban House uses on-site waste recycling, aquaculture, a wind pump, intensive food production, and solar energy for heating and cooking. In 1977, 50,000 visitors toured this center to learn how the new energy techniques worked and to consider how they might improve the ecological character of their own homes. The Farallones Rural Center, on an 80-acre ranch north of San Francisco, is similar to that of the Urban Center, but includes activities more adapted to rural living, such as raising livestock and feed crops and planting orchards. Both centers are firmly committed to being testing grounds and visible examples of *integrated* living systems, which incorporate food production, natural energy systems, and waste reclamation.

The institute offers a large number of educational programs, including year-long resident apprenticeships, regular classes, and one-day workshops. Tours at both the Urban and Rural Centers are open to the public, a significant number of whom are introduced each year to new energy-efficient techniques for daily living.

In its active research programs at the centers, the institute sponsors such projects as the comparison of different passive solar home heating systems, solar greenhouse applications, and the small-scale manufacture of architectural clay products from local materials. The institute publishes the results of these programs. It also offers technical assistance to individuals, to communities, and to government agencies, ranging from the design of skill-development programs and educational materials for the California Conservation Corps to the design and construction of economical, solar-heated residences for individuals. Money realized from publications and consulting fees combine with memberships and private donations to support the institute's work.

The Farallones Institute has a full-time administrative staff, residents, and apprentices, as well as part-time research associates, assistants, and volunteers. The staff represents a great variety of skills and disciplines, from solar design, horticulture and biology, to graphics, publications, bookkeeping, tour guiding, course giving, ceramics, and even blacksmithing.

A similar nonprofit organization, the New Alchemy Institute (founded earlier than Farallones) graces a Cape Cod, Massachusetts site, which is replete with windmills, aquaculture, solar collectors, and greenhouse-living quarters called "arks," ecological micro-systems that combine natural energy sources with fish and plant culture held in symbiotic balance. In 1976 a new ark was opened on Prince Edward Island, with Canadian Prime Minister Pierre Trudeau present for the dedication ceremonies. Other arks are being planned in a variety of climates.

The New Alchemists devote their efforts to monitoring and assessing the arks and to doing research on shelter systems and food and energy production techniques that could be adopted by individuals and small groups. They open the arks to the public, report their findings in journals, and engage in other educational programs. The New Alchemists tend to have professional training—in fields like architecture, engineering, nutrition, horticulture, biology, physics, aeronautics. Their viewpoint is broad, however, and they do not confine themselves to their own disciplines. They are working on an integrated concept, called the family "bioshelter." This is defined as a self-sufficient living environment that could provide basic energy and food requirements for a family without polluting or depleting nonrenewable resources.

The same professional backgrounds are being used in many new, private firms set up to deal with energy-efficient building. Although these places are not experimenting as fully with total life support systems as Farallones or New Alchemy, they recognize that human shelter should not be a separate entity, but rather an extension of the environment. They provide design and consulting services in relation to energy conservation, active and passive solar building, greenhouses, home gardening, composting toilets, and landscaping for wind barriers and summer cooling. Newspaper advertisements or a telephone directory's yellow pages often give the names of architects or building firms that specialize in solar design. These are usually small firms, but they take on additional professional and other help when they expand.

Some of the small firms, like Total Environmental Action, Inc., in southwestern New Hampshire, have gone beyond design and consulting services to become valuable resources for the public and for workers in the field of solar energy. TEA was founded in 1973 by Bruce Anderson, an MIT architect and engineer. Anderson believed the country's move toward efficient use of energy to be essential. His *The Solar Home Book*,[2] is one of the field's most widely read texts. He made a committed statement when starting his own company by retrofitting old, unused mill buildings in a small country town to serve as offices.

People from surrounding communities and from all parts of the United States take workshops at TEA on the fundamentals of solar energy, water power, heating with wood, and food production and preservation. TEA does research for various clients, including the government, for which it also writes proposals. It has a nonprofit foundation and publishes a wide variety of books and visual aids that are low cost, in keeping with Anderson's wish to make such information available to all Americans.

In Maine, two innovative organizations are specializing in helping people build their own homes, or at least to participate fully in their design and construction. Cornerstones in Brunswick, Maine and the Shelter Institute in Bath, Maine attract people from all over the country to two-week seminars in which the fundamentals of building energy-efficient homes are taught. Participants gain practical experience by actually working on homes under construction in the area.

Many of the projects carried on by groups such as these fall into the category of "appropriate technology" a phrase derived from Schumacher's writings.[3] Schumacher argued that we need to make use of the best modern knowledge and experience to develop methods and equipment that are small, decentralized; inexpensive, accessible to all; developed through the use and efforts of local materials and people; and labor-intensive. A central goal of appropriate technology is to increase people's self-reliance at a local or regional level and thereby reduce their dependence on large, centralized systems. The word "appropriate" refers to the suitability of the technology being applied to the local environment. An individual "bioshelter," for instance, might be suited to a rural setting, while a larger, neighborhood unit, utilizing the same techniques in a somewhat different way, would fit an urban scene. Regional climate is also an important factor. Due to the warm climate and the amount of sunlight available during the year, the Southwest leads the nation in potential for using energy from the sun. New England, with less than a third of this potential, needs to back up its use of solar energy with more conventional systems. The kinds of solar energy systems possible also vary according to location. While the Northwest has wood and water power available, the Midwest has more wind. Part of becoming self-reliant will be the use of a great variety of energy-saving techniques that will supplement each other and be appropriate for a specific geographic area.

A National Center for Appropriate Technology, located in Butte, Montana, has been established by the Community Services Administration (CSA), formerly the Office of Economic Opportunity (see Part IV, "Government"). The center, commissioned to develop and to implement appropriate technologies for low-income communities, has set up a number of regional outlets and is funding appropriate technology newsletters and community projects throughout the United States.

The central offices of appropriate technology groups are a potential source of employment for people who know how to write for and produce publications. These jobs are particularly suited to liberal arts graduates who have informed themselves about energy. One of the major tasks of the new groups is to exchange information and to increase the public's participation in conserving energy. Most of these groups maintain libraries or resource centers in addition to disseminating their own publications. Some also do films and TV programs, provide speakers, and hold conferences. People interested in working in all types of media, as well as people who are efficient in office and conference management, sometimes find challenging opportunities here.

Some of the appropriate technology groups also sponsor projects that are staffed by workers in the inner cities and rural poverty areas, who are attempting to respond to the needs of their communities. On East 11th Street in New York City, the Energy Task Force converted two tenement buildings to the use of solar collectors and a windmill. In general, the task force serves as a technical assistance organization, providing community education in simple, low-cost energy technologies. It helped another Lower East Side community group, the Cultural Understanding and Neighborhood Development Organization, install a 534-square-foot solar wall in a previously unheated gymnasium. Neighborhood youths who use the gymnasium constructed the wall themselves. The salaries of many of the people working on local projects like these are funded by government agencies such as the Community Services Administration and the Comprehensive Education and Training Act (CETA). All projects welcome volunteers and have been starting points for many people who want to become involved in the development of solar energy.

Energy Audits and Weatherizing

One person who used volunteer time to advantage is Paul Dryfoos, Energy Technician at the University of Massachusetts Cooperative Extension Service (part of the U.S. Department of Agriculture). After graduating from college with a degree in experimental psychology, Dryfoos did computer programming part-time in order to support himself. He directed his main efforts toward the energy field, working without pay on building projects, with antinuclear organizations, and with the Western Massachusetts Solar Work Group. After two years of experience, he sought out the places that would give him a chance to continue in the field and was successful in getting a job with the Energy Conservation Analysis Project (ECAP), an innovative program at the Extension Service.

In less than a year after its start in 1977, ECAP had performed 2,550 free energy audits for home owners in over half of the counties in Massachusetts. The auditors use standardized engineering techniques to determine heat loss throughout each home and then make recommendations to the owners on preventing the loss. Such recommendations do not include new solar equipment, but are confined to measures like insulating ceilings and walls, tightening up windows and doors, reducing temperatures in hot water storage tanks and putting oil burners in good running condition. Home owners for whom audits have been done often comment on the value of their personal contact with the auditors, who are trained to increase people's awareness of their habits in regard to energy use. The Extension Service hopes to help make major changes in the fuel consumption patterns of home owners in Massachusetts, where, in 1970, the Census Bureau counted 1 million single family dwellings.

The Extension Service employs mostly CETA-funded workers and provides training for a period of 10 classroom days and 10 supervised field days. Trainees learn something about housing construction, heat flow, infiltration, conduction, convection, and combustion, matters that affect heat loss and heat transfer. They receive continued in-service training, some in special aspects of energy conservation. Auditor Deborah Kates has become a wood-heating specialist, working with fire departments and municipalities to inspect wood-stove installation and to help inform the public of safety measures that are necessary when heating with wood.

In addition to running the program for energy auditors, Dryfoos and other administrators are responsible for gathering useful publications for both the office and the public. They handle calls for technical assistance on a problem such as how to decide on the size of a hot water heater or how best to cover a solar collector. They use a combination of practical technical knowledge and administrative skill.

The Extension Service is only one of the places that are developing the new field often called "weatherizing." Others doing even more are the many local Community Action Programs (CAPs) sponsored by the CSA. Unlike the users of ECAP, who come from all income levels and use their own money to implement an auditor's recommendations, the users of CAP weatherizing services must qualify as low-income families and materials and labor will be supplied to them by the agency. People who work as auditors and installers are often funded by CETA, although a few permanent staff positions exist at each agency.

Most of the smaller agencies have only on-the-job training and rely on able supervisors to make their programs work. A few of the larger agencies offer trainees education that includes both theory and practice. In 1978, ECON, Inc. of Boston started a 26-week CETA training program that included lectures, tutorials, and internships. The students were a well-proportioned mix of men and women, whites and minorities, with ages ranging from 17 to 39 years and education levels from 9th grade to college graduation. All but two of the original group of trainees entered employment shortly after completing the course. One woman started her own company to do rehabilitation and energy conservation work.

People who have jobs as coordinators and directors of energy at the local CAPs come from a variety of backgrounds, often having worked as auditors or in other community service positions. These jobs can be frustrating because those who hold them must establish ties with CETA and other agencies in order to obtain funding for their work crews and the federal funding process is often slow and erratic. Coordinators and directors gain a great deal of administrative experience if they can surmount such obstacles and keep the work going on. Crew members learn not only how to make energy audits, but also how to install insulation, storm windows, and weatherstripping, and to repair oil burners.

The State-level weatherizing programs sometimes have sprung from a concern about the high unemployment rate among young people. California established a program to train low-income minority youths as solar technicians and, after an eight-week training period, placed 90 percent of its trainees in private industry. In Michigan, 3,500 youths, including 1,000 who were facing court action, received energy conservation training and summer jobs under the state's youth employment program. Pennsylvania's Department of Community Affairs, using federal and state funds, has run a successful program for several years that has weatherized more than 35,000 homes and trained over 800 people, most of them previously unemployed, in skills ranging from basic construction to energy auditing.

Some community-based programs are being extended beyond conservation measures to include solar systems as well. A new federally funded project, Solar Utilization/Economic Development and Employment (SUEDE), will train CETA workers to design, construct, install, and maintain solar hot water and space heating systems.

Training in weatherization programs can lead to future jobs if a worker knows how to seek out related places. Andrea Addison, who was trained by ECON, went on to work for the nonprofit Boston Architectural Center (BAC), helping to develop its conservation information service for professional builders and architects. Leroy Barros, trained in the same program, moved to working on Boston's housing rehabilitation program. Depending on an individual's interests, motivation, and education level, there are numerous potential employers for those trained in energy auditing and installing:

Fuel companies ▪ oil burner repair; development of auxiliary energy conservation services.

Plumbing companies ▪ installation of solar hot water heaters.

Utilities ▪ energy audits (some utilities now offer this service and all are required to offer it by the provisions in the National Energy Act).

Manufacturers and retailers of building materials ▪ selling or installing insulation, storm windows, weatherstripping; energy audits.

Banks ▪ assessing property that has solar installations; advising on loans for solar systems or weatherization.

Real Estate ▪ selling and buying property that has solar installations or has been retrofitted; advising on weatherization of older homes.

Municipal and county governments ▪ energy audits of government buildings and schools; administering codes requiring weatherization in new buildings; inspection of safety requirements for heating with wood.

Consulting firms ▪ energy audits for industries and private homes.

The state energy offices are using technical assistants, administrators, and publication specialists. These offices apply for and receive the federal funds to be used in weatherization programs. Then, they distribute the funds and monitor the quality of the work of local agencies. They also administer federal and state tax credits for solar installations. Much of their work involves educating the public on how to save energy. People in state energy offices produce advertisements, radio spots, manuals, and other publications dealing with everything from insulating materials and new technologies to energy habits in the home.

New Forms of Education

Not only the state energy offices, but almost all organizations and agencies concerned with solar energy have an educational component. The biggest job is to educate the public about energy efficiency in their homes and daily lives. Education must also be provided to train people like engineers, builders, and installers to handle active and passive solar systems and to keep professionals already in the field up-to-date on new ideas and methods. Then lawyers, real-estate developers, bankers, and insurance agents need to be educated to relate their industries and services to the new technologies.

The education taking place in schools, universities, and community colleges will be covered in following parts of the book. In these settings, most of the educators serve as existing faculty members or administrators of their respective educational institutions and evolve material on solar energy within established departments and curricula. There are many programs going on outside the established educational institutions, however, where innovative ways of reaching out to learners are being tried. It is in these non-traditional settings that some people get started in solar education.

A variety of people and organizations are beginning to get involved in educating the general public about energy. Perhaps the most active and most directly related are those concerned in some way with the environment. The Horticultural Society, Sierra Club, Forestry Association, Audubon Society, and science museums are offering lectures, seminars, short courses, films, slide presentations, and resource libraries on energy topics. From such bases, both paid staff and volunteers can initiate and develop solar education.

Other organizations are to be found in various regional and local areas. Church groups, rotary clubs, women's clubs—almost all organizations that run regular programs for their members—now welcome presentations on solar energy. Although giving such a program for a small group may not be a career in itself, it does constitute an easy introduction to the field and it can lead to other educational jobs, like developing similar programs for college continuing education divisions or for professional organizations (comprised of, for example, bank loan officers or real estate agents).

The Boston Architectural Center is an example of a professional school that added staff members when it created its Service for Energy Conservation in Architecture (SECA). Design professionals and building-related industries use SECA's services, ones that range from technical information and research to sponsoring conferences and reports on legislation and building codes affecting efficient energy use.

The League of Women Voters (LWV) has been especially active in educating the public throughout the country. The Missouri League prepared a 14-minute slide show, with 80 slides and a cassette-recorded script, explaining the importance and potential of solar energy for home owners and businesses. The show includes several examples of Missouri solar projects. A survey booklet of Missouri solar buildings is available. In Maine, the LWV cosponsored with the Audubon Society three separate energy workshops and trade fairs at different locations. The more than 60 exhibits included solar collectors, wind generators, and insulating materials. Workshops covered such topics as home woodlot management and solar hot water heaters. The women who work on planning and implementing LWV programs learn how to put together the topic and the public and, in the process, gain many contacts in the solar field.

Solar fairs and conferences are an important aspect of today's education. They seem to be among the most popular means of increasing the public's awareness of energy options, of helping new energy-related businesses to become visible, and of facilitating shop talk about technical and commercial developments. One of the first large events designed to reach all of these groups was the Toward Tomorrow Fair. It was held for several years at the University of Massachusetts, where it featured exhibits, speakers, workshops, and entertainment, attracting some 15,000 people each day.

In 1978, the Western Regional Solar Heating and Cooling Workshop and Product Exhibition was held in Phoenix, Arizona. This event is an example of the large conference that is aimed at disseminating practical information to professionals. It covered such subjects as building standards and codes, the forecasting of markets for solar products, and the use of government incentives for solar development. Beneficiaries of such a conference include builders and developers, home improvement contractors, plumbing and heating contractors, architects, engineers, utility representatives, lenders and investors, state and local officials, manufacturers, and swimming pool contractors.

How can someone interested in being on the staff of a conference find its sponsors? The first place to look is in the American section of the International Solar Energy Society (see Part IV, "Organizations"). This is the most comprehensive of the solar energy organizations, with members representing all dimensions of the field. Its national and regional chapters usually hold annual conferences, sometimes large in scope so as to reach a wide public audience, sometimes small and technical. This society and the Solar Energy Industries Association are the most regular promotors of events that combine commercial exhibits, speakers, workshops, and seminars.

A few large fairs are promoted by professional convention companies, like Environmental Productions of New England, Inc., in Marblehead, Massachusetts. Their New Earth Exposition, in Boston in 1977, was attended by some 25,000 people coming from a 150-mile radius. Another type of promoter (or, here, sponsor) is an educational institution, most often a college or university.

Working on a fair or conference is a temporary job, but it provides an excellent introduction to the solar education field. Usually such events are announced about a year in advance and their dates can be found by contacting the solar organizations or by scanning their publications. *Solar Age* is the official magazine of the American section of the International Solar Energy Society, and *Solar Engineering* the counterpart of the Solar Energy Industries Association. Extra workers will be hired anywhere from three to six months before a fair or conference. The New England Solar Energy Association added 18 people to their staff for their large BTU Conference in 1977. For a very large conference, the association takes on a director a year or so in advance. Getting to know who some potential employers might be is an added advantage in working on a solar event.

Some universities are committed to serving the larger community through programs that reach beyond the classroom. The University of Vermont has developed an Energy Forum designed to reach a rural population. The Forum gathered audiences in several locations throughout the state to view lectures and demonstrations in Interact TV studios. Those who attended held discussions with one another and with speakers and experts following the presentations. People interested in initiating or working on energy programs can approach educational radio and TV stations, as well as universities.

There is plenty of opportunity to use imagination in bringing solar energy education to all segments of the public. For instance, several energetic young actors, artists, musicians, and educators have formed The Phoenix Nest in South Acworth, New Hampshire. They put on programs in elementary school classrooms (one was called "Sun Celebration") that incorporate several art forms such as mime, puppetry, juggling, song, dance, and mask making. The children play parts in the presentations. Enjoying themselves thoroughly, they hardly realize that they are also learning how energy from the sun can be used.

Another version of using theater and art for education about energy alternatives was developed by a traveling troupe of actors, writers, musicians, and technicians, known as the New Western Energy Show, in Billings, Montana. The troupe has performed the Energy Show innumerable times and now has made available a 40-page book containing the entire script, music, illustrations, and recommendations for stage, set, and costuming (NWES, 226 Power Block, Helena, Montana 59601). People who want to organize informal performing arts groups in their communities could start with this script or create their own presentations. The important topic of energy can give focus to a beginning group.

On the same day in late 1978 one could have watched a 1st-grade class learning about the sun or he or she could have chosen instead to view 11 inmates at the Somers Correctional Institution in Connecticut as they built a 50-gallon solar water heater for the prison. The six-month Somers course, which included instruction in the classroom, as well as learning by doing, led to certification for its graduates and help in job placement when they were released. Two other prisons, in Florida and Tennessee, have started solar technology instruction. The Department of Energy expects some 50 prisons to start offering similar programs.

Teaching, performing, lecturing, and organizing energy education outside the usual classroom setting are all options for those interested in education.

Small Businesses

The belief that America is a nation of entrepreneurs is given credence by the response to the consciousness of solar energy. Untold numbers of independent inventors work to discover better ways of using the sun's rays. Commercial enterprises spring up daily. Some, like the mail order services that offer books on energy topics, are essentially cottage industries and require a minimum of capital. Other, and more visible enterprises are retail stores. A small number are attempting the more ambitious task of manufacturing products used to create energy-efficient shelters. Builders and contractors are, of course, already in small businesses that lend themselves naturally to solar applications.

In climates where a lot of fuel is needed for heating and where wood, that renewable resource, is plentiful, the wood stove business is thriving. Although usually not suited for inner-city use, heating with wood is becoming more and more popular with suburban and rural home owners, and is an example of how a new energy pattern can stimulate a whole new industry. Small manufacturers make the stoves; small retailers sell them. Other manufacturers make chain saws, winches, and axes for the woodcutters, as well as log carriers, shovels, tongs, and an array of accessories. Woodcutters supply and sell the wood. Masons build chimneys. Chimney sweeps clean the chimneys. Publishers print books of instruction. Adult education seminars advise consumers on how to choose among nearly 500 different stoves on the market and on how to cut wood properly from their own property. So the industry goes, on through a chain of accountants, bankers, secretaries, and others who provide auxiliary services.

Already the industry has its own professional organization, the Wood Energy Institute in Camden, Maine, which has publications, seminars, and various projects, such as a national laboratory program to certify the quality and safety of wood-burning appliances. The institute is a good resource for someone interested in starting a wood-heating-connected business.

Some retail stores have begun to specialize in stoves and wood burning accessories, and many small stores that carry hardware or building supplies have added stoves to their wares. Actually, selling stoves is still a business that can be run from home. One family in New Hampshire had been doing car repairs in a large garage attached to their house. The garage is now full of foreign and domestic cast iron stoves for sale. Other dealers have built on showrooms or are using barns or other structures that have the necessary space.

The "energy store" is another new kind of retailing that is appealing to small businesspeople. Such stores carry a variety of items for the consumer interested in energy conservation and the use of renewable resources. Usually they need locations closer to other retail businesses than do the home ventures and therefore require more capital investment. When Buck Robinson started the Cambridge Alternative Power Company, he took over an empty gas station in Cambridge, Massachusetts, and invested $85,000 to renovate and stock it. The store carries everything from water-saving shower heads to solar collectors. The Northeast Carry Trading Company, with the same types of items, operates both a retail outlet and a mail-order business from a Hallowell, Maine, store. Albie Barden, one of the founders of the company, once said that he felt too often people who started energy-related businesses were "cloudy about bill paying, advertising costs, and record keeping," putting excessive attention into the "ideology" of their enterprises. He said he had come to realize that a solid business providing products and services is a way in which people can express their basic philosophical concepts.[4] Energy retail stores can now be found in almost every state in the union.

A number of entrepreneurs have entered the manufacturing field, producing solar collectors, solar-heated domestic hot water systems, storm windows and doors, and insulating materials. Although large, well established corporations are manufacturing solar products, there is also room for the small to medium size business. Kalwall Corporation, in Manchester, New Hampshire, which, a few years ago, had only three people in its Solar Components Division, now has over 30 and exemplifies the successful small manufacturer and distributor. Sensor Technology, Inc., in Chatworth, California, producing solar cells, is still a medium size business, with 250 employees. Anyone considering a manufacturing business will find the booklet, "Starting Your Own Energy Business," very helpful (see Part IV, "Directories").

A potential manufacturer must assess the market or markets for a product. Eventually, it will be sold to one or more of the following: to a retailer; another manufacturer using it as a component; an architect, designer, builder, or contractor, who specifies or uses the product in a particular project.

Important for such a manufacturer is determining ahead of time if he or she has any access to federal grants or private loans—particularly if the product is a new one. The Department of Energy has been running an Appropriate Technology Small Grants Program to support new technologies and processes that maximize the use of renewable energy sources. Small businesses, defined as having 100 employees or less, are eligible for grants to develop and test products or systems in preparation for marketing them.

The Small Business Administration (SBA) is the main source of help for people who do not already have ways of raising money or getting business advice. The SBA advises on drawing up business plans, a first step often skipped by optimists but essential for success in most businesses. The SBA also consults on setting up business procedures, obtaining supplies, and raising money. In 1978 the federal government passed a law providing a loan fund, administered by the SBA, for new companies in the solar and energy conservation fields. Eligible are firms that manufacture, sell, or install energy conservation, solar energy, biomass, industrial cogeneration, hydroelectric, and wind energy products or systems. Also eligible are firms that do design, engineering, and consulting.

Although statistics on solar businesses are not yet available, there is no reason to believe that such businesses are different from other types in their rates of success. The reality is that 80 percent of new businesses end up in failure. SBA studies show that the most frequent reasons for failure are personal differences between principals and lack of cash flow. Still, those who are prepared for and enjoy taking risks and working long hours will tell you that running their own businesses is the only way for them.

Part III

Solar Jobs in Traditional Settings

Private Corporations

"If the nation decides it wants to go solar, we'll be ready to respond," commented a vice-president of General Electric in 1978. Even if solar energy will contribute only a few percentage points of the nation's energy needs by the year 2000, this executive saw it as "a large business possibility" for GE and "something we have to pursue."[1]

Several of the country's largest corporations have come to the same conclusion. Although their investments in solar products and services are still very small, they deem it wise to gain a foothold in the field. Mobil Oil, for instance, while continuing to put its main emphasis on new technologies for mining oil, coal, and uranium, has started a joint venture with Tyco, called Mobil Tyco Solar Energy Corporation, to develop a unique process for manufacturing solar cells. Exxon has invested in an Energy Ventures Division, with one subsidiary developing solar cells and another manufacturing solar collectors and hot water heaters, all being marketed by a new sales staff of nearly 100 people. Honeywell, Inc., not only supplies controls for active solar systems, but is involved in many other facets of the field. In conjunction with Lennox Industries, Inc., Honeywell is actively promoting a domestic hot water system through their more than 6,000 dealers in Canada and the United States. A 1977 survey showed that at least 45 gas and electric companies were already monitoring solar demonstration installations, with 146 companies to be added in 1978.[2] Some utilities foresee owning solar heating, cooling, and hot water systems that they would lease to customers.

In the very large corporations, solar programs are less than 1 percent of their total business, and their staffs are small, usually somewhere between 50 and 500 people. A medium size firm that is devoted solely to solar products actually may employ more people than does a solar project within a larger company. Most jobs fall within four general categories: production, research and development, sales, and administration. The industry, still very young, encompasses everything from the individual innovator and the small business to the corporate giant. Those who prefer to work in the private sector will want to consider the social implications of size, as well as its effect on their working environments.

Some of the people working in the relatively small solar enterprises are concerned that their efforts and expertise will be overtaken by the powerful corporations. The national and international companies with their prestige, their access to capital, and their huge advertising budgets, have tremendous advantages in the marketplace. This, however, does not seem to be inhibiting the start-up of new businesses; over 2,000 companies in the United States are now manufacturing solar products. Solar advocates working in large corporations feel that mass production can bring down costs and make solar systems available to an increasingly large number of people. Many of the things they are working on, like solar cells or rooftop collectors, lend themselves to use right in the home, thereby increasing the self-sufficiency of the individual citizen.

So far, the big, the small, and the medium size businesses are coexisting, their ultimate relationships still to be worked out. Large manufacturers are supplying parts and materials needed by small manufacturers; small manufacturers are supplying components needed by large manufacturers. With rising transportation costs, the manufacture of solar and energy conservation products as close as possible to the point of use becomes more economical with each year. Some products, however (for instance, glass and aluminum), which require very large capital investments, must be made near a particular raw material, and thus continue to be centralized. In general, widespread manufacturing of solar equipment has the potential to move the country toward greater decentralization of energy and commerce, thereby creating a more balanced mix of big and small.

Large companies with solar projects come in a range of sizes: Champion Home Builders Company has 5,200 employees; Grumman Corporation, 28,700; Rockwell International, 115,000; and General Motors, 797,000.[3] From the point of view of the employee, they are all "large." They can be expected to have financial stability and well-developed personnel policies, with good salaries and fringe benefits. In contrast, workers in smaller firms tend to take more financial risk, but they could be part of very rapid expansions within the next decade. They also tend to be more involved in the total operation of the enterprise, in contrast to the specialization of functions usually found in large companies.

So far, the main thrust in the industrial sector has been in the manufacturing of active and passive solar systems for heating buildings and for hot water. The most common component of these systems, at the present state of the technology, has been the rooftop flat-plate collector; as early as 1977, 4 million square feet had been installed in the United States. In addition to the residential market, there is a huge potential market for solar hot water and space heating in large commercial and government buildings.

Relatively sophisticated types of collectors, using evacuated tubes or reflectors to concentrate the sunlight, are also being sold, but are suitable mainly for industrial processes. Anheuser-Busch, Inc., in Jacksonville, Florida, now uses evacuated tube solar collectors to produce hot water for beer pasteurization.

Companies not directly involved with solar projects are at least concerned with finding ways to conserve energy within their own operations. Motivated by rising fuel costs and prodded by the federal government, most large companies are committed to tightening up their buildings and readjusting their manufacturing processes. The 1977 Annual Report of the Celanese Corporation, a chemical manufacturer, stated:

> The commitment of Celanese and its employees to conserving energy and feedstock resources is producing results. By the end of 1977, Celanese had reduced energy consumption by 31.9% in comparison with the base year. . . . Rigorous conservation programs, in effect in all locations, will provide additional energy savings in the future.

Experience seems to be bearing out estimates made by the Ford Foundation[4] that industrial energy use could be cut in half with no major adverse impacts on productivity. In 1976 the John Deere-Davenport Works, of Iowa, needed 32 million BTUs of energy to manufacture one ton of product (commercial loaders and excavators); two years later only 6.7 million BTUs were used per ton. This was accomplished by adding to the manufacturing plant fiberglass insulation, electrostatic air cleaners that allow recirculation of factory air, a central control and monitoring system for heating, ventilating, and air conditioning, and a computer that forecasts electric demand and calculates optimal start-up times.[5]

Companies that do not have industrial processing plants are equally interested in energy used. Sears, Roebuck Company has issued directives to all stores, distribution centers, and office units, that detail for each specific location how the lighting, heating, cooling, and ventilating systems should be regulated to conserve energy. All types of industries need energy specialists to plan their conservation programs and to see that they are carried out.

Large corporations with solar products provide opportunities for people with engineering or technical training. Research is often basic to solar development. A technical background is also valued in production, sales, and administration. A degree from a business school or business experience, especially if it is combined with an understanding of the basic principles of solar or energy conservation, is another way of gaining entry to a professional job. Knowledgeable and enthusiastic sales people, even without formal technical training, are often welcome. Companies with consumer products need field service technicians and those in manufacturing need skilled workers (such as sheet metal workers and assemblers).

The Trades

Some of the unions are actively advocating the development of solar energy. In March 1978, the Energy Subcommittee of the Congressional Joint Economic Committee held hearings on "Creating Jobs Through Energy Policy." Subcommittee Chairman Senator Edward Kennedy (D-MA) heard testimony from union leaders who felt that national energy policy and full employment were related and should be solved together.[6] They argued that a vigorous conservation-plus-solar energy course for the nation was the way to create more jobs.

Several in-depth studies have been done and show that using renewable energy sources is extremely labor-intensive and creates far more jobs than the nuclear or any traditional sources.[7] Skilled workers will be in increasing demand as solar energy and conservation gain momentum.

People in all the building trades already are involved in building energy-efficient homes and retrofitting older ones. In climates where solar heating and cooling systems are backed up by conventional systems, the maintainance of two systems creates additional work. The installation of storm windows and doors, insulation, weatherstripping, and other conservation measures used in passive solar heating, in both old and new homes, also creates jobs. Since increase in the energy-efficiency of buildings—both public and private—is of primary importance, the largest growth in the trades is taking place in the installation, maintainance, and repair of solar and energy conservation systems.

The second largest growth is in the making of the products consumers use. Machinists, assemblers, and almost all kinds of production workers are employed by companies turning out collectors and other components, as well as insulating materials.

At one time it was assumed that if the United States were to use less energy, unemployment would follow. Yet it was known that West Germany, Sweden, and Switzerland, with standards of living comparable to those of the United States and low rates of unemployment, used only about one-half the energy per person used in this country. Recently, the assumption that a large increase in our capacity to generate electricity is our only path to economic growth, jobs, and the "good life" has been challenged. During 1977, the State of California had a 50 percent decline in its electricity growth rate and at the same time personal income rose 12.5 percent and nearly 500,000 new jobs were created.[8] The experience of companies that have decreased their energy consumption without decreasing production or employment, along with well-documented studies, add evidence that the conventional wisdom about the relation between energy growth and employment is becoming outdated.

Significantly, more and more unions are embracing the solar-conservation combination. People in the trades, both union and nonunion, who turn toward these fields will find opportunities increasing every year.

Government

The Northeast Solar Energy Center in Cambridge, Massachusetts, is operated by the Northern Energy Corporation, a nonprofit organization, and it is funded by the U.S. Department of Energy, putting it in a category somewhere in between private and government enterprise. It is one of four regional centers formed to foster the commercialization of solar energy. This center covers the New England states, and New Jersey, New York, and Pennsylvania.

Steven Brown, a young member of the staff, went to the center not just because he was offered a good job but also because he considers himself an advocate of solar energy. After getting his BA in economics from Trinity College, he worked as a product manager for a manufacturer in Bridgeport, Connecticut. When he read Schumacher's *Small is Beautiful*,[9] he started thinking about how he could further this philosophy in his own work. It took him almost two years to integrate his thoughts and set his priorities. He came to the conclusion that efficient transportation and renewable energy sources were the most important developments necessary to sustain a society running out of fossil fuels, and he decided that he would try to find a way to earn a living in one of these fields.

Brown's first interest was in the increased use of bicycles for commuting and regular transportation. He conducted an extensive study of utility bicycling and its effect on energy conservation and pollution abatement, but he was unsuccessful in finding a way to bring his ideas into the commercial marketplace. Turning to solar energy, Brown approached a small manufacturer of solar hot water systems who needed improved marketing techniques and offered his services. He went to work for this firm, becoming responsible for sales, public relations, and training distributors and sales personnel. This experience enabled him to take the next step to the Northeast Solar Energy Center.

At the center, Brown serves as a liaison with the technical community, often going into the field to determine the needs for training programs and then following up with implementation and funding from the center. He has worked on developing installer training programs, passive solar design seminars for architects, solar energy courses for public sector personnel, and other technical training courses. He says that he sees a big difference between people who are solar advocates and those who are just filling a job, both at the center and in the field. He particularly enjoys the opportunity his job gives him to associate with solar advocates in all parts of the Northeast.

Brown remains an avid bicyclist and has managed to retain some of the activities he enjoyed before moving to a more urban area. Although he prefers the kind of personal life he led in a small town, he was willing to make the change for the kind of work he can do at the center. He feels that people who really want to work in the solar field can find a way, but they must be willing to take time and to gain preliminary experience that will be useful.

Most of the people working at the center have had experience in other fields and have chosen to bring this experience to bear on the development of renewable energy resources. They come from jobs in industrial, architectural, construction, academic, research, financial, legal, real estate, and insurance firms. Their educational backgrounds range from liberal arts degrees to highly technical and professional training in engineering, business, and the sciences.

Although the center still has some of the informality of a new organization, in many ways it does not appear on the surface to be much different from the offices of the companies with which the employees were formerly involved. There are reports to be written, telephone calls coming in, meetings taking place. The people are engaged in contacts far beyond the office, however, and must constantly devise ways to break new ground. This creates a variety in the working day not often found in more established industries.

"I have to use a cross-section of skills unlike anything I have known before," said Joseph Levangie, Manager of the Industrial Development Division. Levangie, a chemical engineer with an MBA from the Harvard Business School, has worked for other companies where he was responsible for transferring new business ventures from the research stage to the marketplace. In 1975, he wrote a report on energy for Representative Paul Tsongas (now Senator, D-MA) and became seriously interested in the solar field. Levangie lives in a suburb, and, like most of the other people working in urban locations of the solar industry, leads a life similar to that of his peers in other industries, but he and his colleagues tend to have an added sense of urgency about the work they are doing.

The four regional Solar Energy Centers sponsored by the Department of Energy are like most government agencies in that they deal in services rather than in products. People who enjoy administrative work can look for opportunities administering the solar programs that are run by government departments or funded by government money. The "Government" section in Part IV lists some of the state and federal government offices that are involved with solar projects and energy conservation. Although all of these offices need staff, it is helpful to be aware of a few general trends in government solar activity.

One way in which the government is encouraging the growth of solar systems and conservation is by providing *financial incentives such as tax credits and small grants* for businesspeople and home owners who install renewable energy systems and add insulating features. Loans also are made to new businesses in these fields. Some people who work on this kind of federal program are located in Washington, D.C., but since most of the money is channeled through state agencies, many more work in their state capitols. Some states have enacted their own grants and tax incentive programs.

Another source of government employment is in the area of *demonstration projects on the use of solar energy in buildings*. By 1978, over $6 million in solar grants had been awarded by the Department of Housing and Urban Affairs for residential unit construction. The Department of Energy helps finance many types of solar installations. The world's largest hospital solar heating unit, for instance, was constructed on the roof and terraces of the Danbury Hospital, in Connecticut, with the projection that over $35,000 in oil and electric costs will be saved each year. The Department of Energy, which provided two-thirds of the financing, worked with two private firms, Arthur D. Little, Inc., of Cambridge, Massachusetts, and Shepley Bulfinch Richardson & Abbott, architects of Boston, on the technical aspects of the project. In developing these commercial and institutional uses, people in government agencies often work closely with community development commissions, housing authorities, and private developers.

A third area of growth in government employment involves *making government property energy-efficient.* The massive heat loss in public buildings was ignored in the years when energy was cheap, but now government at all levels—federal, state, county, and municipal—is busy trying to lower their fuel bills. Some states have special funds and staffs available to do energy audits of public school buildings and county units are often active in energy planning. The county planning office in Hampshire County, Massachusetts, has done a study of how public buildings can decrease their energy use dramatically by lowering the temperature of hot water, using off-peak electric rates, turning off hot water heaters by timers when the buildings are closed, and adding insulation. This study recommends including solar energy in all new buildings and suggests a new solar-heated House of Correction.

Meanwhile, in neighboring Franklin County, Massachusetts, a recently completed energy project has projected the county's energy picture in the year 2000, comparing renewable and nonrenewable sources. The Department of Energy has awarded the project a $30,000 grant to develop a community education program and to determine what renewable energy programs can be implemented. Similar programs are going on in other parts of the United States. People who would like to work in planning for conservation and the use of solar energy would do well to look into government activity at the county level.

An increase in the use of solar energy in government buildings has been stimulated by the passage of the Hart Amendment, sponsored by Senator Gary Hart (D-CO), to the 1979 Military Construction Act, which provides for $100 million to be spent annually for solar installations on military housing and other buildings. People in the services now will have a chance to get training and experience as solar technicians and the increased demand for solar products will help the commercial development of low-cost systems.

Some people will want to consider another new kind of government job, although there will be only a relatively small number of opportunities. This is the job of *legislative energy aide.* Many United States senators and representatives now have energy aides on their staffs and legislative committees involved in energy matters have the same. Some young people got into these jobs after working in support of specific bills or on an event such as Sun Day, where they came in contact with legislators and learned the politics of energy. The work involves keeping track of all energy developments, both in and out of the government, preparing position papers, answering questions from constituents, attending hearings and conferences, and in general being the "energy expert" in an office.

One large government program stands in a class by itself. The Tennessee Valley Authority (TVA), under the leadership of David S. Freeman, Chairman of the Board of Directors, has made an active commitment to solar development, with a goal of having renewable resources provide more than one-third of the valley's energy by 1990, and a majority of it by 2000. Among other projects, TVA plans to make available technical and financial assistance for the installation of 1,000 solar water heating systems, to be expanded throughout the region from its Memphis site, and for the construction of passive solar homes in a resettlement community for people moving from a flood plain area.

Administration in a large government solar program can be an extremely challenging job. As Director of Solar Outreach and Technology for TVA, architect Elizabeth Chase provides the technical backup for professionals in the private sector who are designing and building solar structures. She is responsible for providing architects, engineers, contractors, and developers with technical seminars, assistance with design, and the use of sophisticated computer facilities. She shows her enthusiasm when she explains what is being accomplished through the TVA program: "We're involved with the construction of a Chattanooga office building of 2 million square feet that will use only 30,000 BTUs per square foot a year, compared to 115–150,000 BTUs per square foot in a typical office building. We're making extensive use of natural lighting, using waste heat from the computer, ground water cooling, energy zoning—there are so many things that are possible to do."

Chase came to government work after more than 15 years in the academic world as a researcher, writer, editor, and teacher. She said that her first interest in solar energy was sparked several years ago by one of her students who was working on a solar project. In regard to her present job, she says, "I was wanting to get right into the middle of the action, and I did."

Anyone starting to explore government work should contact the nearest Civil Service office and find out what kinds of jobs require taking the Civil Service examination. Highly professional specialists, such as lawyers and architects, are usually exempt. Many other people would do well to take the exam and get a state and-or federal Civil Service rating, which places them on the available list and facilitates matters greatly if a job opportunity occurs.

Those who know of specific government solar programs that are of interest to them can get in touch with their senators or representatives and ask for help in making contacts in the appropriate departments. For state jobs, state senators or representatives are the ones to see. Helping constituents through the bureaucracy is one of the functions of an elected representative.

Research

Although government and industry are already developing energy conservation and solar construction, using methods that are well known, an extended use of solar energy will be based on many new technologies. It has been estimated that enough sunlight falls on the earth each day to satisfy mankind's energy need for 15 years. How to contain that sunlight and convert it to usable energy, and how then to keep from wasting it, is a challenge that has attracted a growing number of people who are interested in scientific and engineering research.

Research projects can range from exploring the most basic physical principles—such as the nature of the energy contained in the tides or in the heat at the core of the earth—to developing technical devices needed to use the sun's energy in everyday life. Although they all could affect the efficient use of energy in buildings, some projects are directly related to this problem. Experimenting with, testing, and developing better solar panels, hot water systems, insulating materials, and building methods all fall in this category.

Other kinds of projects are less obvious to the aspiring researcher. One of these was carried out by the Energy Research and Development Administration (ERDA), now the Department of Energy, and the Potomac Electric Power Company, a Washington, DC, based electric utility. Seventy homes were equipped with a special device that gives up-to-the-minute readings on the cost of electricity being used at any time during the day. The device, about six inches square and two inches thick, can be placed in a kitchen or other convenient location so that consumers can tell at a glance how much the cost will be of continuing to use the amount of electricity being used at the moment during the next hour. They can see immediately the cost difference when large energy-using equipment, such as air conditioners, are started up. The research was designed to answer an important question: will consumers cut down on their energy use if they are reminded regularly what it costs?

This particular project is a good example of the way research in the solar and conservation fields can utilize many different kinds of skills and organizations. Involved in this study were an independent inventor, a university, private industry, and government.

The monitoring device, a Fitch Energy Monitor, was developed by R. B. Fitch, a Chapel Hill, North Carolina, builder who has incorporated energy-saving features into the homes he builds. The project evolved from research conducted for ERDA by Princeton Univesity, where students read meters daily and passed the information on to home owners. (The result was a 10 percent reduction in electricity use.) Princeton researchers continued to provide research and evaluation support for the project. The Potomac Electric Power Company was a joint sponsor and provided 140 homes (70 to have monitors and 70 as a control group) from a cross-section of the utility's service area. They installed the devices and kept track of electricity consumption. Dr. Maxine Savitz, the Director of the Division of Buildings and Community Systems in the Department of Energy, initiated the study. Most of the division's efforts to encourage energy efficiency had focused on developing new products and technologies; Dr. Savitz wanted to reach the consumer directly. "Even after they get their bills, consumers are not able to see which equipment uses the most power," she said.

Although solar research topics cover psychology, economics, sociology, and most other social science disciplines, much of the research going on deals with the technology of solar energy. A number of people are working in wind energy. The development of solar cells, or photovoltaics, is another promising technology in need of scientists and engineers.

Solar cells generate electricity directly from the sun by creating a current when the light hits certain sensitive materials. Cells can be made from various materials and in several different structures. Advances in the technology are coming rapidly, but an efficient cell is still extremely costly to make. As more markets develop, stimulating mass production, cost is expected to drop drastically. Some experts have compared the position of the solar cell today to the position of the transistor just before it became available in the mass market at very low cost.

Transistors brought an almost revolutionary change to many consumer products and manufacturing processes; solar cells could have an even more significant effect. Some envision solar cells on every roof top, thereby giving people a measure of energy independence they have not had since electricity became a household necessity. Jeold Noel, an Oregon solid state physicist, has estimated that the roof of an average house could produce enough energy to supply the needs of the home, with enough energy left over to charge an electric car.[10] Some utility companies, on the other hand, are predicting that solar cells will be used in centralized power plants generating electricity to be sold to consumers.

Solar cells already are being used in satellites and spacecraft, on off-shore oil platforms, for remote telephone and television transmission stations, on Coast Guard buoys, and to power radios, lights, signs, refrigerators, irrigation units, and other systems, usually in remote locations where electrical power is not available. The commercial use of photovoltaics in buildings is just beginning. A ten-unit condominium is under construction in Palo Alto, California, and a $1.1 million solar cell powered classroom facility is underway at Mississippi County Community College in Arkansas.

The Solar Energy Research Institute (SERI), in Golden, Colorado, has classified the major solar technologies now under development in the United States as: solar heating and cooling of buildings; agricultural and industrial process heat applications; solar thermal electric generation; photovoltaics; wind energy conversion; bioconversion; ocean thermal energy conversion; and hydroelectric power.[11] Some of these technologies are maturing rapidly, while others still require essential research, development, and demonstration. In all cases, more research is needed on their influence on economic and social conditions and on how public policy should be used in their development.

SERI, operated for the Department of Energy by the Midwest Research Institute, has a mandate from Congress to develop solar technologies and to help establish an industrial base that will lead to widespread use of solar energy. Therefore, its research includes both scientific-engineering and economic-social science disciplines. It is the largest group ever assembled for work devoted exclusively to the development of solar energy; plans call for an organization of over 700 professional and support personnel when it is fully staffed.

In the technical field, SERI has established its own research programs in photovoltaics, biomass energy, and industrial process heat, and has undertaken management of several Department of Energy contracts in other solar technologies. Less than half of SERI's staff is composed of scientists and engineers. The rest are sociologists, economists, political scientists, information specialists, market analysts, lawyers, architects, and other professionals. These are the people who research the consumer, commercial, industrial, and political issues, and help the government develop national solar energy programs and strategies.

They also provide educational programs and data about solar energy, not only to the technical community, but also to federal, state, and local offices and legislative bodies, and to manufacturers, architects, the financial community, and home owners. They plan to have a computerized Solar Energy Information Data Bank operating by the early 1980s.

Large corporations, government agencies, and universities are the usual settings for basic technological and economic research on solar energy. Researchers often move back and forth between them once the researchers have gained a detailed knowledge of a particular aspect of their work. Mechanical engineering, architecture, electrical engineering, physics, and chemistry are the disciplines most frequently used in technical research.

Most researchers make their initial job contacts while they are doing their graduate study, leading them to the place that is working on problems of most interest to them. Relying on colleagues and the communication network within the sciences, including professional journals, is an effective way to get a start on a career. The best way to find out about many aspects of the field is to read the two main solar periodicals, *Solar Age* and *Solar Engineering Magazine*.[12]

Public Schools and Colleges

If the adult's urge to do research often has its beginning in early childhood, our supply of solar scientists may be adequate in the future. Nobody knows how many small children recently rigged up two coffee cans full of water, one painted black and one painted white, and set them on a sunny windowsill with a thermometer in each to monitor the temperatures. In school science fairs, students are eager to display everything from special lenses focusing the sun's rays to ignite paper or wood to solar cookers and model flat plate solar collectors. The subject of solar energy and energy conservation fascinates many young people, and, starting at the elementary school level, teachers and administrators have an opportunity to participate.

Since, at almost every school, the curriculum already is crowded with material, some educators choose to encourage students' energy education through extracurricular means, such as the science fairs mentioned above, or the Energy Conservation Corps created by Richard Ammentorp in Schaumburg, Illinois, for 3rd-through 6th-graders. The club, which meets after school, influences the entire school and community. It calls attention to energy waste, issues Energy Saver Certificates to home-rooms, conducts energy poster contests, and writes and stages energy plays.

The main thrust of energy education in the schools, however, can be seen in the integration of new material into existing curricula. On the secondary level, such material appears in science classes, but also in social studies or multidisciplinary units that have been developed for use in several subjects. It covers a wide spectrum of energy sources, both traditional and alternative, and considers also the political and social implications of their use.

Energy material is readily available to teachers and is proliferating every day. A solar energy curriculum for use in grades 1 through 12 has been developed by two Department of Energy sponsored projects,[13] and many state education departments, professional organizations, community groups, and publishing companies have prepared energy materials for school use. For a survey of this information, see *The Energy Education Materials Inventory,*[14] which lists and describes material put out by all these sources, and includes things like films and games, as well as books, workbooks, classroom activity units, and teachers' manuals.

The National Science Teachers' Association has established the Task Force for an Energy-enriched Curriculum, which keeps up with new developments and publishes a free newsletter, "Energy and Education." The Spring issue carries a list of two-to four-week summer institutes on energy for high school and college teachers and administrators. In 1979, the Department of Energy funded 66 such institutes in all parts of the United States, many of them specializing in energy conservation, solar and other energy alternatives.

Some school districts have their own summer workshops to develop energy education. An added interest can result when teachers are involved in creating the program and when students have an opportunity to relate the material to their own particular location and climate. In addition to a knowledge of larger problems, they need to understand the ways in which they can affect their own communities.

Although all educators cannot initiate energy education programs in their schools, they can get ideas from comprehensive programs, such as the school-based Staples Energy Conservation Program in Minnesota. Largely funded by the U.S. Office of Education (Title IV), this program was started by a committee of 18 Staples teachers from every subject and grade level and was designed to reach all the students in both the public and parochial schools, their families, the towns, and the surrounding farm communities. It provides not only classroom material for all teachers but also a large resource center outside the school.

One example of the program's influence occurred when a local church borrowed a heat scanner from the Staples center and discovered that a costly new insulating job had been done improperly. Many individuals, businesses, and farms are now installing improved insulation in old buildings, using passive solar design in new ones, starting to drive small cars, and changing their habits to be more conserving. The students are engaged in studies and projects that have a visible connection to the "real world"; the goal in Staples is to have energy efficiency both taught and practiced.

Educators who start energy programs in their own schools and find that they would like to do more might consider becoming associated with one of the commercial, nonprofit, or government agencies that disseminate energy education materials. State departments of education in California, Iowa, Maryland, Minnesota, and New York (and several other states) are accumulating educational materials and helping local school districts develop energy curricula. Publishers, nonprofit, and grass roots organizations are calling upon educators to help produce films, filmstrips, slides, activity projects, and written material.

In higher education, the opportunities for incorporating solar energy in the curriculum are more oriented toward technical matters, largely because research and graduate training tend to be the focus of financial support coming from private industry and the government. Sometimes it is difficult to coordinate people working on technology with those interested in the technology's social and political aspects. At other times, new energy studies are not available at the undergraduate level. Nevertheless, a number of faculty members and administrators have been successful in initiating courses related to alternative energy and in some instances also have started up impressive energy centers with services available to the public as well.

There is an unwritten expectation that higher educational institutions supported by public funds will be resources for help in solving some of the problems faced by society at large. The public colleges, universities, and community colleges appear to be responding to this expectation. They are in the forefront of developing solar research, engineering, undergraduate courses, and education for the general public. Many community colleges are training skilled installation and maintenance people for work on solar equipment.

Educators at all levels can become further involved in solar energy by attending or helping to organize conferences that deal with energy. In 1979 a major conference sponsored by the Department of Energy attracted 1,300 representatives from education, business, and labor to Washington, DC, to discuss mutual concerns and goals connected with the future of energy in the United States. In a "Turning on Energy Education" conference, also in 1979, the Los Angeles City-County Energy Education Council gathered together education leaders, teachers, and administrators to exchange energy knowledge and skills and to coordinate energy education efforts. Through professional organizations and government units, educators interested in energy are being offered increasing opportunities to reach out beyond their classrooms and to work with others with similar concerns.

Communications

The appearance of educational materials on energy in the nation's classrooms is only one small part of the ferment going on in energy information. The "energy crisis" was once perceived as a distant problem to be dealt with somehow by officials in Washington. Now, confronted with the Three Mile Island nuclear plant accident in Harrisburg, Pennsylvania, and local shortages of gasoline, the American people feel personally involved and are personally absorbing information that formerly had been of only slight interest to them.

The TV news has a special segment on the scientists working on solar cells; the local newspaper features a nearby home that has a solar hot water heater; the radio carries the proceedings of a solar conference. The growing coverage of renewable energy sources means that anyone already working in the established communications systems has an opportunity to carve out a specialty. Although the by-line that reads "Steven Rattner reports on energy for *The New York Times*" gets more national attention, many regional and local newspapers also have energy reporters. The specialty usually includes coverage of all energy sources, of which solar is one.

The usual journalistic methods of tracking down information through reading, conferences, and talking to people in the field can be preceded by and supplemented with courses or seminars in the basic principles of solar energy. Once started, the journalist will find so many leads that it will be hard to know which ones to follow. For those who are not yet part of the communications field, specializing in the subject of energy is one way of gaining entry. Some people have started with free-lance articles in local newspapers—weeklies, dailies, Sunday supplements, shoppers' newspapers—or with helping to produce local radio or cable TV programs. With samples of their work on renewable energy in hand, they can then approach a larger publication or station.

The field of communications includes varied professional people, such as writers, editors, film makers, radio and TV writers and producers, graphic artists, copy writers, photographers, educators, and public relations specialists. Any of these people can turn their attention and expertise toward solar energy, either by introducing the subject into their existing jobs or by finding new jobs in which it exists already.

Newspapers, radio, and TV are the most visible segments of the communications field but many other places are also providing solar information. Some of them, because they are reaching highly specialized audiences, are more effective than the mass media in getting their material read or heard.

Some potential employers, or potential recipients of free-lance work, are:

Professional and trade associations ▪ publish newspapers, journals, magazines, and reports to inform members of solar developments; some use brochures, films, and exhibits.

Nonprofit community organizations and university-based energy centers ▪ usually circulate newsletters and large numbers of short reports and other publications.

Magazines ▪ articles on solar energy are used by magazines of general interest and by magazines oriented toward science, the consumer, or the home; free-lance writers will also find a market in magazines that specialize in the alternative energy field (for instance, *Solar Age, Solar Engineering Magazine, Co-Evolution Quarterly, Rain, Alternative Sources of Energy Magazine*).

Book publishers, audiovisual publishers, film producers ▪ print materials on solar energy, especially when their publishing specialty is education or science and technology.

Private corporations, with solar products or services ▪ need technical manuals and films, brochures, exhibits for fairs and conferences, advertising copy, company newsletters, government proposals.

Government (federal, state, and local) ▪ uses newsletters, reports, press releases, brochures, flyers, conference exhibits, advertising copy (e.g., advising people to conserve energy), and research reports, particularly in energy and energy-related departments, such as housing and agriculture.

Needless to say, engineers, installers, and others who are already employed in the technical area of the solar field and *also* can write or produce visual materials are very much in demand.

Related Fields Needing Solar Specialists

The fields of education and communication account for only some of the demand for people who can inform the public about solar energy. The use of solar systems in buildings already has had an impact on the related commercial institutions, creating opportunities for people in banking, insurance, real estate, investment, and the law. It would be misleading, however, to suggest that specific jobs are waiting for the solar advocate; instead, people in these fields are finding ways in which their knowledge of solar energy can be used as part of their regular responsibilities.

How does a bank loan officer, when asked for mortgage money, assess a house or commercial building that has solar heating or hot water systems? This may be the most common solar business transaction to date for both business and nonbusiness people. Unfortunately, many local bankers have little knowledge of what constitutes an efficient solar system and so may approach the matter with suspicion.

Some lending institutions have hired outside firms as professional consultants to answer the above question. Others have tended to rely on the brand names of the large national manufacturers, a practice that may have an adverse effect on the success of local or small businesses with quality products. As more and more buildings make use of solar energy, banks are beginning to need in-house personnel who can make well-informed judgments on these new technologies.

In 1979 the Department of Energy ran a series of solar workshops for the financial community in cities across the country. Many state and local energy agencies, trade associations, and nonprofit organizations have run similar sessions in their own regions. Interested bankers can find ways in which to learn about solar technologies, and, in addition to using that learning in their own business dealings, can help to educate their colleagues.

Growing numbers of banking firms are providing financial assistance to customers who want to "go solar"; some even offer slightly lower interest rates on solar and energy conservation loans. A few banks, found in all parts of the country, have active or passive solar heating or solar hot water systems in their own buildings—an excellent way for their personnel to become knowledgeable. Commercial banks, savings-and-loan associations, and credit unions all require people who can respond when the lending institution is asked to finance an individual or a developer who is using solar equipment in new buildings, to make loans for retrofitting older buildings, or to provide working capital and other financial services for solar equipment manufacturers, research and development firms, architects, and other businesses in solar energy.

The same need is felt in the insurance industry. At an insurance workshop held by the Northeast Solar Energy Center,[15] participants agreed that they should have little difficulty with providing product liability insurance for the components of solar systems. The pipes, valves, hot water tanks, and glass used in these systems have been found in conventional systems and generally around the home for years. The problem felt by the participants was in the underwriters' and insurers' lack of knowledge about solar energy, which may lead them to refuse insurance for products without knowing that there are no new or unusual risks involved.

The workshop participants did express some hesitancy in extending liability insurance to owners of completely installed solar energy systems. There is no way of telling whether or not a system has been installed properly. Improper installation by an unknowledgeable contractor or a "do-it-yourselfer" could result in failure or unexpected damage.

This doubt can be resolved in part by insurance people who take the time to find out which contractors and installers have the competence that warrants confidence. The existence of standards for products and services would be an even more useful aid. Various trade and solar industry associations are working to define standards by means of which particular products and systems can be tested and certified. They are also developing some sort of certification for workers who do solar installation. Solar-energy-related standards gradually are being incorporated into local building codes so that building inspectors can attest that a solar system is of a certain quality.

Real estate people, whether they are brokers or developers, have similar concerns in relation to lack of knowledge about and standards for solar energy systems. A local real estate salesperson, in showing a house with active or passive solar features, often has little appreciation of the quality of the solar work or of the possible solar benefits. An understanding that a lot of insulation is a good selling point is about as far as some salespeople can go.

A developer of residential or commercial property has an even more complex job. Should the new houses have solar hot water heaters? Should the new office building be heated partially by wood and municipal wastes? The options are many. The developers must be knowledgeable enough to employ the most competent designers and to make the right decisions.

In relating their fields to solar energy, people in real estate, insurance, and banking will be responsible for answers to questions such as the following:[16]

Is the initial cost of the solar energy system fair and realistic?

How long will it take to pay back the initial investment?

What would be the replacement cost of the system?

How reliable is the system?

Does it perform efficiently?

How long will it last?

Are people available who are qualified to install and maintain it?

What is the impact of the solar energy system on the building's insurability and resale value and on obtaining a second mortgage?

What local, state, and federal regulations affect the system?

How susceptible is it to any damage that may be caused by weather, fire, water, vandalism, electrical malfunctioning?

Sometimes, there will be additional questions concerning the manufacture and sale of solar products. Moving into solar in these settings will appeal to people who enjoy the interface between technology and commerce.

Becoming a solar stock specialist is another solar job possibility for people in the field of finance. Many solar product and services companies offer stock to the public. Institutions that have large investments, investment firms, mutual funds companies, and stock brokerage houses are some of the places where knowledge about solar energy firms, and about solar energy in general, is needed. *Solar Engineering* carries a monthly report on solar stocks written by specialist Anthony W. Adler.

Lawyers who represent solar businesses or clients with solar systems may find themselves dealing with many of the financial and technical issues discussed above, but some of these issues relate specifically to the law.

The one that has generated the most interest concerns the protection of access to the sun for a building's solar collectors. They can be rendered useless if neighboring land owners build high buildings, grow big trees, erect large fences, or otherwise block the lateral sunlight. Prospective buyers of solar equipment should know what, if any, their legal rights are in the event that there is any attempted or actual blockage of the sunlight on which that equipment depends.

Another legal issue involves zoning regulations, which are the province of local governments. Such regulations may pose many obstacles to the use of solar energy. In upstate New York, for instance, a landowner was ordered by the police to remove a solar collector from his front yard, and, when he refused to do so, he was taken to court. Zoning regulations that cover such matters as heights of buildings, lot boundaries, out-buildings, additions to existing buildings, and the use of residential property to produce electric power, as well as aesthetic and architectural controls, all affect the use of solar energy.

A third issue arises when building codes do not provide for solar energy systems, and local officials have discretionary power. They may deny a building permit, or, in some cases, require costly testing of the design or strength of the system before issuing a permit.

A growing and swiftly changing body of law is developing to govern the increasing use of solar energy. Each solar technology has its own particular legal problems, and each state and locality has its own regulations affecting the technologies. Some regulations, usually the outdated ones, inhibit solar use. Others encourage it. Legislation in San Diego County, California, requires solar hot water systems to be installed in new residential construction in subdivisions located where electricity is the only alternative; in a later phase it is planned to extend this regulation to areas where natural gas is available.

The Department of Energy's Solar Energy Research Institute (SERI) publishes *The Solar Law Reporter*,[17] a bimonthly that contains information on pending and adopted statutes, regulations, and judicial decisions, and analyzes legal problems and issues. The American Bar Foundation conducted a study and published the results in *Overcoming Legal Uncertainties About Use of Solar Energy Systems*.[18] In addition to exploring legal problems, that study suggests ways in which federal, state, and local governments can help implement the development of solar energy.

Lawyers may represent clients on issues involving the use of solar energy in buildings, or, when business clients, on ones involving patents, product liability, and warranties for solar products and services. Also, lawyers have a special function in helping to develop policies and practices that promote solar development, without impinging on individual rights or on the welfare of the community. The full use of solar energy in buildings cannot take place without updated building codes, revised zoning regulations, and new laws that govern access to the sun.

Part IV

Resources for Information, Jobs, and Education

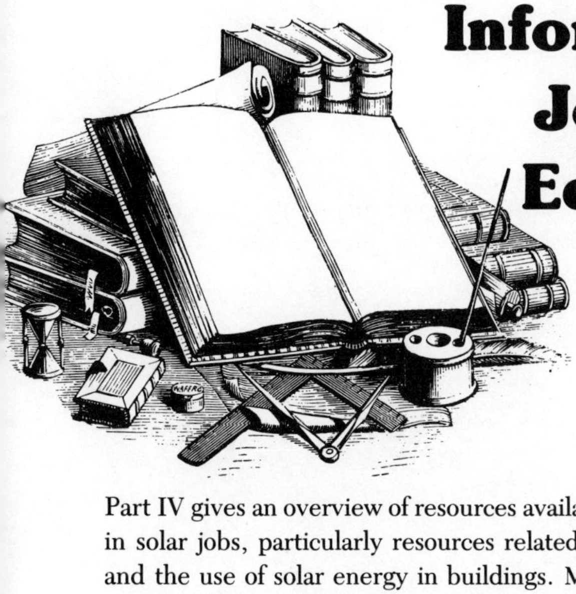

Part IV gives an overview of resources available to people interested in solar jobs, particularly resources related to energy conservation and the use of solar energy in buildings. Material gathered from a number of sources has been coordinated and summarized so that it can be presented in concise form. For more detailed information, readers should write or call the agencies, institutions, and organizations in which they are interested.

How to Get Started

If you are just beginning to get interested in working in the solar energy field, the most important first step is to inform yourself—a step that costs little or nothing and may well lead to further steps for you. At least it will reveal the extent of your interest before you make a commitment. Some of the ways of exploring the field are:

1. Seek out people near you who are active in the field, talk with them to find out what they and their organizations do. People engaged in developing solar energy—in industry, government, universities, and other organizations—usually are enthusiastic about their work and are glad to talk to someone who has a genuine interest.

2. If you are a student, line up a term paper, an internship, or a summer job in the field. If a student in the sciences or engineering, consult with faculty members and read the professional journals to find out where the best solar research and development projects are going on. Highly trained technical people have many opportunities open to them.

3. Take a short seminar on solar energy. Not only will you learn some of the basic principles and applications, but also you will meet other people with similar interests.

4. Read in the library or send for a few publications that interest you. *Solar Age* and *Solar Engineering Magazine*, the two main solar periodicals, will give you a view of current activities in the field (see footnote 12, Part III).

5. If already working, see if you can get involved in energy concerns right where you are. For instance, you could try to introduce energy conservation measures in your building, to add a section on solar energy to your company library, or to work on including some energy items in a product line. Meanwhile, you can explore the ways in which your present experience can relate to going on more seriously in the field.

6. Get involved as a volunteer with a community organization or professional society.

7. If still planning your education, consider some of the college programs that specialize in solar energy or alternative technologies.

The best career decisions are often made after a period of time during which a person mixes exploration, working (paid or unpaid), and training. In a new, rapidly expanding field, it is wise to remain flexible enough to move back and forth among these activities.

Once you have made a decision to look for a job, two questions must be answered: Do I want to stay in my present location or am I willing to go to other parts of the country? and Do I already have a skill or profession that I want to use or am I open to a new kind of occupation? Many people feel somewhat ambivalent about the answers to these questions, but job hunting cannot be very effective unless they are answered, at least tentatively. You do, after all, have to start looking somewhere. And you will need to state to prospective employers what you think you can do for them.

Organizations and businesses involved in the development of solar energy report that the subject seems to interest many people whom they interview about jobs. When asked, "What can you do?,"more often than not the job seeker answers something like, "Well, I don't *do* anything special, but I'm willing to learn. I'm really interested–and here I am!"

Consider some replies that would be more appealing:

> I understand that your organization puts out several publications. I have taken some short seminars on solar energy and have done a lot of reading on the subject. I'd be interested in writing reports and newsletters or doing other kinds of public relations work for you.
>
> I've trained and worked as a computer programmer. I feel that the development of solar energy is very important and would like to be a part of your research project. I think I'd want to get involved beyond computer work eventually.
>
> I want to get started in the solar building field and to learn carpentry and masonry. Can you use me on your crew?
>
> Now that your department is handling tax credits, funding of demonstration projects, and so many aspects of conservation and solar energy, I thought you might be looking for some additional administrative staff people. I have experience in administering a

small social agency, I've informed myself about solar energy and I'd like to get involved in its development.

I feel your organization is doing very important work and I want to work with you. In college I helped plan and carry out a lot of student activities. If you need help, I think I might be good at helping in the office and organizing your large meetings and conferences.

You may not start out doing any of the above, but your chances of finding a job will be a hundred times greater if you have some notion of what you want to do. Furthermore, do your "homework"—research each potential place of employment prior to an interview. Starting with these things in your head—some notion about what you want to do and some research—your discussion with an employer can lead to your identifying a range of tasks wide enough so that you may agree upon a job.

A helpful book on exploring and job hunting is Richard Bolles' *What Color Is Your Parachute?* (Ten Speed Press, Berkeley, CA 1979). It stresses the importance of individual initiative in finding compatible work, an initiative essential in getting started in the solar energy field. Then, of course, there are always the classified ads for job openings, but, as in any field, probably the most interesting opportunities will not surface there.

Directories

This section can be used as a source of ideas for work in the solar energy field. The listings include specific geographic locations, which should facilitate inquiries about job opportunities. Public libraries can be requested to buy the directories or they can be ordered directly from the publishers. All are paperbacks.

The Solar Engineering Master Catalog and Solar Industry Index, 1980. $15. Published by Solar Engineering Magazine, 8435 North Stemmons Freeway, Dallas, TX 75247. Information on companies throughout the United States involved with all types of solar energy systems—a comprehensive guide to solar products and the industry. Lists solar manufacturers, installers, maintenance contractors, architects, engineers, computer services, research consultants, and so forth.

Solar Energy Information Locator, 1978. Free. Published by the Solar Energy Research Institute (established by the U.S. Department of Energy), 1536 Cole Boulevard, Golden, CO 80401. A small, easy-to-use booklet prepared from the latest information from the Solar Energy Information Data Bank being developed by the institute. Gives addresses and phones, areas of interest, publications, and information services for organizations and government agencies.

Citizens Energy Directory, 1978. $7.50. Published by Citizens Energy Project, 1413 K Street NW, 8th Floor, Washington, DC 20005. Listings by states of nonprofit organizations, government agencies, citizens' action and community groups, solar energy newsletters and magazines, educational and research groups, and some private companies. Notes areas of interest, services available, publications, and contact people.

Starting Your Own Energy Business, 1978. $4. Published by Institute for Local Self-Reliance, 1717 8th Street NW, Washington, DC 20009. Provides practical information on how to start a small retail or manufacturing business in four fields: retrofitting, storm windows and doors, cellulose insulation, and solar hot water systems. Emphasis is on enterprises requiring relatively low capital investment. Lists Small Business Administration field offices in all states.

Energy: A Guide to Organizations and Information Resources in the United States, Second Edition, 1978. Published by Public Affairs Clearinghouse, P.O. Box 10, Claremont, CA 91711. Descriptive listing of government agencies and other organizations, divided into sections according to type of energy. Includes all energy sources—oil, gas, coal, nuclear, electric utilities, and so on, as well as alternative sources such as solar, wind, and water.

1977 Solar Energy & Research Directory. $24. Published by Ann Arbor Science Publishers, Inc., P.O. Box 1425, Ann Arbor, MI 48106. Listing of individuals, companies, and research facilities, divided according to type of activity (that is, manufacturing, distribution, design, research) and cross-referenced when more than one activity is involved. Although some listings are now out of date, the research section carries more entries than other directories and indicates many of the places where solar research is well established.

Northeast Yellow Pages of Solar Energy Resources, 1978. $4. Published by New England Solar Energy Association, P.O. Box 451, Brattleboro, VT 05301. A directory of solar and other alternative energy companies that offer products and services available to New Jersey, New York, Pennsylvania, and New England consumers. Other regional Solar Energy Society offices (see list of organization, p. 78) are preparing directories for other parts of the country.

The Solar Age Resource Book, 1979. $9.95. Published by Everst House, N.Y. Includes 16 articles by solar experts, a detailed listing of solar equipment and systems available on the market, a geographical directory of architectural and design services, and buyers' guides for wind products, wood products, and all types of solar products.

Informal Directory of the Organizations and People Involved in the Solar Heating of Buildings, Third and Final Edition, 1977. Published by William A. Shurcliff, 19 Appleton Street, Cambridge, MA 02138. The author, a research physicist at Harvard University, has been working in the field of radiation physics for 40 years. This directory, which he put out himself in order to "promote the solar heating of houses, schools, and commercial buildings," was long considered the definitive one. The field has grown considerably since the publication of the final edition, but this directory still has the broadest scope of all, including government agencies, commercial concerns, universities, professional societies, and foundations, as well as the names of individual solar architects, engineers, consultants, inventors, promoters, writers, economists, and home owners.

Directory of State Government Energy-Related Agencies, 1975. $8. Prepared by Federal Energy Administration. Order from National Technical Information Service, U.S. Department of Commerce, 5285 Port Royal Road, Springfield, VA 22161 (publication number PB-246 891). Gives functions, major programs, publications, services, budget, staff size, and names of key personnel in state energy offices. Also includes functions and addresses of other offices, such as energy-related committees of the legislatures, and state contacts for environmental and public utility issues.

National Solar Energy Education Directory, First Edition, 1979. $4.75. Available from Superintendent of Documents, U.S. Government Printing Office, Washington, DC 20402 (stock #061-000-000210-3). The 300-page directory, compiled by the Solar Energy Research Institute (SERI), lists solar-related courses, programs, and curricula at 700 postsecondary institutions nationwide. Detailed information covers address and telephone number of an institution, credits offered, duration of a course or program, topics covered, instructor's name, average enrollment. The information, now stored in a computerized data base, was gathered in the main by responses to a survey.

Organizations

Professional, trade, and other nonprofit organizations are useful sources of general information about solar activity in a particular region or in a special trade or profession. For someone who has not had much exposure to the solar field, making contact with a few of these organizations is one way to start interacting with the people and finding out how to become further involved.

Many people have started working with solar organizations on a volunteer basis by helping to research and write reports, to run workshops and conferences, to publish newsletters, to do general office work, or, in the case of trade organizations, by serving on committees that set standards and plan training for solar technology. If extra help is not needed at the moment, most organizations welcome someone who will gather information on a special topic for future use. Experience and knowledge gained in this way are assets in finding solar-related jobs. And the settings offer opportunities to get to know people in technical, commercial and noncommercial solar enterprises and thus to hear of job openings.

People can also look for paid jobs with nonprofit organizations. Although the staffs are often small, openings do occur as workers move on to other settings.

The reader is reminded to look in the "Government" listings (p. 81), particularly at the state energy offices and the federal Regional Solar Development Centers, National Center for Appropriate Technology, and Community Services Administration as further sources of information on solar activity in their regions.

The following list of organizations, presented by state, is not exhaustive, but should provide some starting points for interested people in most parts of the United States. Each organization may know of others in their area that have inadvertently been omitted here.

United States

ALABAMA

Alabama Solar Energy Association, c/o Dr. G. R. Gunn, Johnson Environmental & Energy Studies, University of Alabama, P.O. Box 1247, Huntsville, AL 35807. Regional chapter of American Section, International Solar Energy Society (see p. 78).

ARIZONA

Arizona Community Action Association Energy Project, 1820 West Washington Street, Suite 203, Phoenix, AZ 85007. Acts as advocate for low-income people; membership includes individuals and community action agencies throughout the state; concerned with demonstration projects, solar workshops, films, resource lists, children's literature on energy, directories, displays at fairs, technical and economic studies.

Arizona Solar Energy Association, c/o Dan Halacy, Arizona State University, Tempe, AZ 85281. Regional chapter of American Section, International Solar Energy Society (see p. 78).

ARKANSAS

Arkansas Consumer Research, 1852 Cross Street, Little Rock, AR 72206. Consumer group disseminating alternative energy information; runs conferences; engages in political action.

MO-ARK Solar Energy Society, c/o Wendall Locke, P.O. Box 401, Jefferson City, MO 65101. Regional chapter of American Section, International Solar Energy Society (see p. 78).

The Ozark Institute, Box 549, Eureka Springs, AR 72632. Center for rural concerns (regional affiliate of the National Center for Appropriate Technology); conducts research, demonstration and training in solar greenhouses; offers technical and information services to regional community action agencies and organizations.

CALIFORNIA

Alternative Energy Collective, 2600 Dwight Way, Room 204, Berkeley, CA 94704. Provides demonstration projects at fairs, solar classes and workshops, technical studies, assistance with solar building; newsletter.

California Public Policy Center, 304 South Broadway, Suite 224, Los Angeles, CA 90013. Carries out studies and planning; publishes reports; promotes solar energy as a means to create jobs.

Citizens for Energy Conservation and Solar Development, Inc., P.O. Box 49173, Los Angeles, CA 90049. Nationwide volunteer organization to carry out projects on conservation and solar greenhouses.

Citizens Home Energy Conservation Project, 1460 Koll Circle, San Jose, CA 95112. Community action group serving senior citizens and low-income families in Santa Clara County with insulation, weatherstripping, furnace filters; community-wide education on energy conservation.

Electric and Gas Industries Association, 1335 Market Street, Suite 411, San Francisco, CA 94103. Solar Division has membership of manufacturers, distributors, retailers, contractors, installers, architects, and designers; provides information on solar-related legislation, technology, commercialization, tax credits, government regulations and codes; puts on educational programs for members and the public; newsletter, pamphlets, and other publications on solar energy and energy conservation.

Farallones Institute, 15290 Coleman Valley Road, Occidental CA 95465; or Integral Urban House, 1516 5th Street, Berkeley, CA 94710. See Text, p. 20.

Marin Citizens for Energy Planning, 80 Lomita Drive, Mill Valley, CA 94910. Community group researching energy issues; occupied with library, workshops, speakers, surveys, reports; newsletter.

Mid-Peninsula Conversion Project, 867 West Dana, No. 203, Mountain View, CA 94041. Community group planning ways in which workers could transfer from military to socially useful production, such as solar and alternative energy, in Santa Clara Valley, where over 100,000 workers rely on military production for jobs; published 80-page report, "Creating Solar Jobs: Options for Military Workers and Communities" and produced slide show, "Swords into Ploughshares"; newsletter.

National Center for Appropriate Technology, Field Representative: David Colfax, The Mountain School, P.O. Box 246, Boonville, CA 95415. See "Government" listings, p. 87

Northern California Solar Energy Association, c/o Dr. Donald W. Aitken, Center for Solar Energy Applications, San Jose State University, San Jose, CA 95192. Regional chapter of American Section, International Solar Energy Society (see p. 78).

Santa Cruz Alternative Energy Co-op, P.O. Box 66959, Scotts Valley, CA 95066. Community group promoting application of alternative energy sources by studies, workshops, speakers, demonstration projects; newsletter.

Southern California Solar Energy Association, c/o John T. Brand, 202 C Street, 11B, San Diego, CA 92101. Regional chapter of American Section, International Solar Energy Society (see p. 78).

Sunrae (Solar Use Now for Resources and Employment), 5679 Hollister Avenue, Room 5B, Goleta, CA 93017. Solar advocacy group supports and helps write laws such as, solar tax

credit; educates the public through workshops, speakers, audiovisual shows, solar information; newsletter. Locations of other offices:
53 Canyon Road, Berkeley, CA 94704; 1107 Ninth Street, Room 1023, Sacramento, CA 95814; 6378 Jeff Street, San Diego, CA 92115; 440 North First Street, San Jose, CA 95112; 1720 Pacific Avenue, No. 311, Venice, CA 90291

COLORADO

Boulder Energy Conservation Center, 929 Pearl Street, Boulder, CO 80302. Center provides services related to energy conservation and low-technology and solar projects; maintains resource material and speakers bureau; runs fairs; workshops include building solar units, retrofitting, greenhouses, passive design, landscaping, and tours of solar projects.

Colorado Solar Energy Association, c/o Rachel Snyder, P.O. Box 5272, TA, Denver, CO 80217. Regional chapter of American Section, International Solar Energy Society (see p. 78).

Educational Media Center, University of Colorado, Stadium 360, Boulder, CO 80302. University-based group producing materials on alternative energy, including films, filmstrips, and cassette programs with accompanying study and teachers guides; also provides conferences, courses, and informal consulting to educate the public.

Environmental Action of Colorado, 2239 East Colfax Avenue, Denver, CO 80206. Center managed by Colorado Solar Energy Association is clearinghouse for information related to state solar energy activities; Solar Bookstore is open six days a week.

National Center for Appropriate Technology, Field Representative: Doug Baston, Rte. 2, Box 725C, Coal Creek Canyon, Golden, CO 80401. See "Government" listings, p. 87.

Pike's Peak Solar Energy Association, P.O. Box 4321, Colorado Springs, CO 80930. Community organization of consumers and technical people promoting conservation, solar, and other appropriate technologies; monthly informational meetings, tours of projects, hands-on workshops on building and installing solar collectors; maintains energy information booths at parks and fairs; newsletter.

San Luis Valley Solar Energy Association, P.O. Box 1284, Alamosa, CO 81101. Community organization promoting development of solar and other alternative resources through locally based efforts in this rural area; monthly informational meetings, tours of solar projects, workshops on building solar systems, speakers bureau; produced directory and slide show of solar people and projects in the valley; newsletter.

Solar Energy Association of Northeastern Colorado, Inc., P.O. Box 307, Eaton, CO 80615. Citizens' educational organization run by volunteers to support use of solar energy; monthly informational meetings, Sun Day exhibits; newsletter.

CONNECTICUT

National Center for Appropriate Technology Field Representative: Robert Shortreed, 84 Westford Road, Stafford Springs, CT 06076. See "Government" listings, p. 87.

People's Action for Clean Energy, P.O. Box 563, Middletown, CT 06457. Citizens' group with six local chapters has literature, films, workshops, exhibits, library, speakers on conservation of energy, alternative energy, and dangers of nuclear pollution.

Solar Association of Connecticut, c/o K. Ramon, P.O. Box 541, Hartford, CT 06101; and 20-A Sycamore Lane, Manchester, CT 06040. Chapter of New England Solar Energy Association-American Section, International Solar Energy Society (see p. 78).

DELAWARE

Delawareans for Energy Conservation, 104 Cambridge Drive, Wilmington, DE 19803. Citizen action group to promote solar and energy conservation throughout the state; provide slide programs for use in schools and in adult community; develop and distribute energy curriculum to public and private schools; sponsor fairs; serve on Governor's Energy Commission and Public Service Commission; newsletter and solar flyer.

DISTRICT OF COLUMBIA

American Institute of Architects, 1735 New York Avenue, Washington, DC 20006. Professional society, with regional chapters and 31,000 members, provides workshops, speakers, and design centers; does continuing education on energy efficiency through design, solar, and other alternative energy systems; publishes "Energy Notebook" series and other publications; public education campaigns and lobbying for energy conservation legislation.

Bureau of Standards, Center for Building Technology, National Engineering Laboratory, Washington, DC 20234. Develops and implements standards for solar heating and cooling equipment and systems.

Center for Renewable Resources and Solar Lobby, 1001 Connecticut Avenue NW, Washington, DC 20036. Citizens' group, an outgrowth of Sun Day, formed to lobby for solar energy and to build a grassroots network around the country to advance decentralized solar technologies; working on survey of solar and energy conservation projects in the United States; monthly magazine, *Sun Times*, and other publications.

Citizens Energy Project, 1413 K Street NW, 8th Floor, Washington, DC 20005. Keeps abreast of all developments in energy field—energy policy at state and national levels, new community projects, publications, individuals with new ideas, resource organizations, and so on; publishes books, pamphlets, manuals, and directories.

Concern, Inc., 2233 Wisconsin Avenue NW, Washington, DC 20007. Research organization, national membership, with demonstration projects, consulting, literature on energy and conservation, quarterly nutrition newsletter; promoted government-sponsored Solar Energy Exhibit; consultants to Department of Energy.

Council of American Buildings Officials, 560 Georgetown Building, 2233 Wisconsin Avenue NW, Washington, DC 20007. Writes building codes that include solar systems.

Institute for Local Self-Reliance, 1717 8th Street NW, Washington, DC 20009. Committed to developing community self-reliance, particularly in urban areas; stresses research, demonstration projects, technical assistance, and review of public policy; supports energy projects such as rooftop greenhouses for low-income senior citizens and energy task forces for conservation and use of renewable energy; publishes 16-page newsletter-magazine that reports almost all new small-scale technologies and programs for urban decentralization.

Interfaith Coalition on Energy, 1413 K Street NW, 8th Floor, Washington, DC 20005. Coalition of five Catholic, Jewish, and Protestant national organizations with campaign to enlist members of churches and synagogues in a "Covenant for Conservation," believing that excessive consumption and waste of energy by Americans is "a cause of social injustice at home and abroad"; provides specific conservation actions for individuals, sermon outlines, and material for educational programs and communal celebrations.

Minority Energy Technical Assistance Program, Center for Urban Environmental Studies, 1012 14th Street NW, Suite 706, Washington, DC 20005. Project to inform minority legislators about energy-related matters through literature, questionnaires, referral to resource organizations in their communities, and training sessions.

League of Women Voters' Educational Fund, 1730 M Street NW, Washington, DC 20036. Pilot programs for energy education in locations throughout the United States reach the local level, stress practical applications of conservation and solar energy, and work on identifying obstcles to energy-efficient technology; programs provide discussion groups and presentations for other local organizations.

Public Interest Research Group, P.O. Box 19312, Washington, DC 20036. Research arm of Ralph Nader organization; has main interest in nuclear power safety and alternative sources of energy; puts out many publications with solar information; some local chapters act on behalf of use of renewable energy sources.

National Association of Plumbing, Heating, and Cooling Contractors, 1016 Twentieth Street NW, Washington, DC 20036. Trade association with strong interest in solar energy and energy conservation; provides literature, seminars, program of energy audits; is writing sections on solar energy into apprentice training; participates in developing voluntary solar contractor certification.

National Association of Solar Contractors, 910 Seventeenth Street NW, suite 928, Washington, DC 20006. Trade association for solar contractors; provides technical information on solar installations and servicing, assistance with sales and marketing, help in finding skilled tradespeople and professionals; represents members in government programs and in supporting legislation; runs workshops; publishes bimonthly technical newsletter and other publications.

National Science Teachers Association, 1742 Connecticut Avenue NW, Washington, DC 20009. Project for Energy-Enriched Curriculum has developed 23 classroom units in energy; teachers participate in three regional conferences to refine and revise the material; newsletter, "Energy and Education."

Public Interest Economics Foundation, 1714 Massachusetts Avenue NW, Washington, DC 20036. Community group promoting energy conservation through educational programs; clearinghouse for professional advice, lobbying; does economic studies, some under government contract.

Solar Energy Industries Association, 1001 Connecticut Avenue NW, Suite 632, Washington, DC 20036. Trade association established in 1974 to promote interests of organizations and individuals involved in commercial advancement of solar energy; represents companies doing research, manufacturing, and installations of solar systems in legislation and regulation affecting the solar industry, in establishing standards for equipment, in gathering statistical and market data, and so forth; keeps members informed of developments in technology, markets, government activity through conferences, exhibits, workshops, and publications; publishes magazine, *Solar Engineering*.

Worldwatch Institute, 1776 Massachusetts Avenue NW, Washington, DC 20036. Independent, nonprofit research organization does research and analysis and publishes findings to bring global issues, including those concerned with energy, to the attention of the public.

FLORIDA

Clean Energy Research Institute, P.O. Box 248294, Coral Gables, FL 33124. Does research in alternative energy; runs short courses; organizes world and national conferences; puts out papers and proceedings.

Environmental Information Center, Florida Conservation Foundation, 935 Orange Avenue, Winter Park, FL 32789. Community group doing research; providing general public education for the state through conferences, research reports; newsletter and special publications on solar applications.

GEORGIA

Georgia Conservancy-Southern Rural Action, 3110 Maple Drive, Suite 407, Atlanta, GA 30305. Citizens' group interested in alternative energy; publishes newsletter funded by National Center for Appropriate Technology.

Georgia Solar Energy Association, c/o Des Yawn, Campus Box 32743, Georgia Institute of Technology, Atlanta, GA 30332. Regional chapter of American Section, International Solar Energy Society (see p. 78).

National Center for Appropriate Technology Field Representative: Tyrone Brooks, 1315 Beecher Street, Atlanta, GA 30310. See "Government" listings, p. 87.

IDAHO

Community Action Agency, Coeur d'Alene Multipurpose Center, P.O. Box 1300, Coeur d'Alene, ID 83814. Community group helping to lessen impact of energy crisis by winterizing homes and experimenting with alternative sources of energy.

ILLINOIS

American Bar Foundation, 1155 East 60th Street, Chicago, IL 60637. Legal research organization; did study of legal problems such as sun rights and of federal legislation related to solar heating.

Illinois Environmental Council, 407½ East Adams Street, Springfield, IL 62701. Coalition of environmental groups lobbies for state legislation on environment-related issues, including energy; newsletter, "IEC News."

Human Environment Planning, Governor's State University, Park Forest South, IL 60466. Nonprofit organization within university; dispenses information on research, development and use of energy alternatives and appropriate technology by people in mid-West; provides speakers, media presentations, reference collection, conference coordination; magazine, *Outlook*.

National Center for Appropriate Technology Field Representative: Jim Laukes, 105 Wolpers Road, Park Forest, IL 60466. See "Government" listings, p. 87.

Northern Illinois Solar Energy Association, c/o J. A. Hartley, P.O. Box 352, Argonne National Laboratory, Argonne, IL 60439. Regional chapter of American Section, International Solar Energy Society (see p. 78).

Pollution and Environmental Problems, Inc., Box 309, Palatine, IL 60067. Volunteer group promoting solar, wind and conservation through energy meetings, solar forums; newsletter.

Shelter Lab One, 7347 South Paxton Street, Chicago, IL 60649. Citizen volunteers run experimental projects on "ecologically viable" shelter (such as aquaculture, construction of geodesic dome); do studies of energy used in construction and running of average American home.

Solar Project, Department of Physical Sciences, Sangamon State University, Springfield, IL 62703. University-based group with demonstration projects, workshops, energy program for high school teachers, speakers, exhibits, films; does consulting.

South Central Illinois Solar Energy Association, c/o Earl G. Powell, P.O. Box 516, Hillsboro, IL 62049. Regional chapter of American Section, International Solar Energy Society (see p. 78).

INDIANA

Alternative Technologies Association, Inc., P.O. Box 20571, Indianapolis, IN 46220. Small nonprofit corporation experimenting with low-cost, labor-intensive projects (has constructed a DHW solar collector and retrofitted a home with passive solar system); sponsors library, workshops, speakers, consulting; newsletter (in conjunction with Breakthrough Plus 40).

IOWA

Community Action Research of Iowa, Inc., P.O. Box 1232, Ames, IA 50010. Nonprofit citizen consulting and advocacy group to encourage utility rate reform and use of conservation and renewable energy resources in mid-West; drafts legislation, litigates, advises citizens' groups, runs conferences, lobbies; "New Criteria," the Midwest Journal of Appropriate Technology, funded by National Center for Appropriate Technology.

KANSAS

Kansas Solar Energy Society, c/o Donald R. Stewart, 1202 South Washington, Wichita, KS 67211. Regional chapter of American Section, International Solar Energy Society (see p. 78). The Kansas Society has eight other chapters, as follows:

K. C. Metro Chapter, c/o Ron Wontoch, 6500 West 69th, Overland Park, KS 66204

Manhattan-Junction City Chapter, c/o John Selfridge, 1509 Houston, Manhattan, KS 66502

Northwest Chapter, c/o Dr. Richard Hayter, 634 Harrison, Suite B, Topeka, KS 66603

South Central Chapter, c/o Steve Burr, Route 2, Box 50 F, Salina, KS 67401

Southeast Chapter, c/o Charles Booth, 1149 North Terrace, Wichita, KS 67208

Southwest Chapter, c/o Gerald Hundley, 508 Stoeckly, Garden City, KS 67846

Topeka-Lawrence Chapter, c/o Kevin Halbach, 1266 Buchanan, Topeka, KS 66604

West Central Chapter, c/o Joseph Pajor, 1602 South McLean Boulevard, Wichita, KS 67213

KENTUCKY

Appalachia-Science in the Public Interest, P.O. Box 612, Corbin, KY 40701. Citizen group provides building technology resource center for Central Appalachia; researches strip-and coal-mining issues; gathers information on solar and conservation matters; provides workshops, speakers, conferences, research reports and other publications; new Solar Energy-Appropriate Technology Demonstration and Research Center under construction.

MAINE

Maine Audubon Society Alternative Energy Network, 118 Old Route 1, Falmouth, ME 04105. Energy project has public education programs, curriculum development; headquarters of society completely heated by solar and wood; publications.

Maine Chapter of New England Solar Energy Association, c/o John Germond, Monmouth, ME 04281; also 24 Goff Street, Auburn, ME 04210. Local chapter of American Section, International Solar Energy Society (see p. 78)

Wood Energy Institute, Camden, ME 04843. See text, p. 31.

MARYLAND

Foundation for Self-Sufficiency, Inc., Central Maryland Foundation Research Center, 35 Maple Avenue, Catonsville, MD 21228. Volunteer organization, with speakers, slide shows, informal consulting, demonstration projects (has own windmills and solar collectors); research in appropriate technology, with findings available to the public; newsletter and other publications; branch offices in Lily Pons and Arnold; memberships by individuals encouraged.

National Solar Heating and Cooling Information Center, P.O. Box 1607, Rockville, MD 20850. See "Government" listings, p. 88.

Volunteers in Technical Assistance (VITA), 3706 Rhode Island Avenue, Mt. Ranier, MD 20822. Private voluntary organization; gathers resources on alternative energy and technologies geared to self-sufficient life styles; disseminates information through handbooks, manuals, technical bulletins, and newsletter to people working in development of Third World countries.

MASSACHUSETTS

Center for Energy Studies, Boston University, 195 Bay State Road, Boston, MA 02215. University-based group with library, demonstration projects, courses, technical studies, speakers, and symposia to promote conservation and solar development in community at large.

Office to Coordinate Energy Research and Education, A225 Graduate Research Center, University of Massachusetts, Amherst, MA 01003. University-sponsored office to foster energy research and educational activities within the university and the state; assists faculty with research funding; facilitates educational programs; provides information for faculty, students, and the general public; assisted in formation of Energy Conservation Analysis Project, program for installation of passive solar equipment in low-income homes; offers series of teacher workshops on energy, several community college faculty development institutes, a journal of appropriate technology, a guidebook to appropriate technology sites in New England.

League of Women Voters of Massachusetts, 120 Boylston Street, Boston, MA 02116. Carries on research studies, workshops, conferences, and lobbying for energy conservation and the development of alternative energy resources.

Mass Bay Chapter of New England Solar Energy Association, c/o Robert O. Smith, 55 Chester Street, Newton, MA 02161. Local chapter of American Section, International Solar Energy Society (see p. 78).

Massachusetts Solar Action Group, 133 Paul Gore Street, Jamaica Plain, MA 02130. Community group; has many members from recently discontinued Massachusetts Solar Action Office, introducing weatherizing and low-cost solar technologies; educates public on energy alternatives; assists with finding funding for solar projects.

Metropolitan Energy Workshop, 74 Joy Street, Boston, MA 02215. Information center providing library, local energy information exchange, speakers, consulting.

New Alchemy Institute, P.O. Box 432, Woods Hole, MA 02543. See text, p. 21.

New England Coastal Power Show, c/o Steve Crowley, 40 Kinnaird Street, Cambridge, MA 02139. Large solar demonstration van, owned by Clamshell Alliance, visits high schools, colleges, fairs, and public meetings; offers workshops and public discussions; demonstrates solar ovens, parabolic troughs and heaters, window-box heaters, photovoltaic cells running small TV sets, and so on; carries literature, posters.

Northeast Appropriate Technology Network, Inc., P.O. Box 548, Greenfield, MA 01301. Promotes research, development, and use of energy alternatives by people in Northeast; keeps up on activities in energy conservation and solar; serves as a clearinghouse for appropriate technology information; publishes magazine, *New Roots*.

Urban Solar Energy Association of Boston, 21 Burnside Avenue, Somerville, MA 02144. Community group; sponsors solar greenhouse "barnraisings," compiles literature on lost-cost retrofitting, runs program monitoring performance of local solar installations; newsletter.

Western Massachusetts Solar Energy Association, c/o Paul Dryfoos or Bob Schrader, Energy Conservation Program, Cooperative Extension Service, Tillson Farm, Amherst, MA 01003. Local chapter of New England Solar Energy Association-American Section, International Solar Energy Society (see p. 78).

MICHIGAN

Alliance for Clean Energy, c/o Rod Bailey, Grand Valley College, Allendale, MI 49401. Citizen action group; organizes workshops and conferences on energy alternatives and neighborhood self-sufficiency.

Ecology Center of Ann Arbor, 417 Detroit Street, Ann Arbor, MI 48104. Community group running recycling station, home heat conservation workshops, demonstrations; developed energy curriculum grades 7–12; has environmental library; newsletter, "Energy Series," *Walker's Guide*, and River Valley Publications.

Energy and Environment Information Project, Wayne State University, 4866 Third Street, Detroit, MI 48202. Community resource group providing information on impact of different types of energy, economic considerations, cost and supply alternatives; newsletter.

Michigan Solar Energy Association, c/o Edward J. Kelly, Jr., 201 East Liberty Street, Ann Arbor, MI 48104. Regional chapter of American Section, International Solar Energy Society (see p. 78).

Project Entropy, Math-Science Teaching Center, Michigan State University, East

Lansing, MI 48824. University-sponsored group runs regional teacher training workshops, young people's energy conferences; has energy demonstration van with displays, multimedia programs, other materials of use to educators; informal consultation with educators; newsletter.

Upland Hills Ecological Awareness Center, 2575 Indian Lake Road, Oxford, MI 48051. Operates Farm School with solar collector heating, wind-generated electricity; gives exhibits and demonstrations, two-to-five and 10-day workshops for participants living on the farm and working on projects; new solar heated, wind powered, sod-roofed, energy-independent building opened in Fall 1979; provides library, speakers, consulting; newsletter.

MINNESOTA

Alternative Sources of Energy Magazine, 105 South Central Avenue, Milaca, MN 56353. Bimonthly magazine covering wide range of alternative energy topics; also sponsors displays at fairs, workshops, library, speakers, energy information, consulting.

Ouroboros South Project, Continuing Education in the Arts, University of Minnesota, 320 Wesbrook Hall, Minneapolis, MN 55455. Information center on design and energy conservation; provides tours of model house, literature on solar energy, wood, wind, methane.

Ramsey Action Energy Conservation, 509 Sibley, St. Paul, MN 55101. Local agency for Ramsey County residents; supported by federal, state, and local funds; runs weatherization program for low-income home owners-renters; does Research Program on Optimum Weatherization, crisis intervention for winter-related energy emergencies, education program for public on energy conservation.

Staples Regional Energy Information Center, Chicago Avenue and 5th Street, Staples, MN 56479. Government-funded educational group serving six-county area developed energy conservation curriculum for grades K–12; serves as community-school resource center for conservation and energy alternatives; sponsors community and school energy fairs; provides library, workshops.

MISSISSIPPI

Institute of Environmental Science, University of Southern Mississippi, Box 314 Southern Station, Hattiesburg, MS 39401. University program to do research and to train environmental scientists; also provides community with speakers, technical and economic studies, demonstration projects.

MISSOURI

Missouri Solar Energy Associates, 2008 Engineering, University of Missouri, Columbia, MO 65201. Maintains library; keeps members informed of solar energy developments and energy problems in the state; newsletter.

MO—ARK Solar Energy Association, c/o Wendall Locke, P.O. Box 401, Jefferson City, MO 65101. Regional chapter of American Section, International Solar Energy Society (see p. 78).

National Center for Appropriate Technology Field Representative: Council Smith, 438 North Skinker, St. Louis, MO 63130. See "Government" listings, p. 87.

MONTANA

Alternative Energy Resources Organization, 435 Stapleton Building, Billings, MT 59101. Tours state with renewable energy show (see text, p. 30); prepares educational presentations, curricula for schools, educational radio programs, children's coloring and cut-out energy books; helps with state solar planning; newsletter.

National Center for Appropriate Technology, P.O. Box 3838, Butte, MT 59701. Headquarters for NCAT. See "Government" listings, p. 87.

NEBRASKA

Greater Omaha Community Action, Inc., 1805 Harney Street, Omaha, NE 68102. Local agency, federally funded; utilizes unemployed adults and youths from low-income population in employment and training program; constructs passive solar window heaters, prefab greenhouses, and other solar devices in ongoing energy conservation and weatherizing program for elderly, handicapped, low-income families in four-county area; maintains library, demonstration projects, speakers' bureau, consulting services; runs energy workshops.

Midwest Energy Alternatives Coalition, P.O. Box 83202, Lincoln, NE 68501. Community group; acts as clearinghouse for publications and activities in solar and energy conservation in Nebraska, Kansas, Iowa, and Missouri.

Midwest Energy Alternatives, 2444 B Street, Lincoln, NE 68502. Community group; has library, speakers, hands-on solar workshops, displays, solar and earth-sheltered house tours, conferences, seminars; newsletter, "MEA Journal."

Small Farm Energy Project, Center for Rural Affairs, P.O. Box 736, Hartington, NE 68739. Research and demonstration project; monitors conservation techniques and renewable energy innovations on 25 low-income farms in Cedar County; measures economic savings; helps farmers become energy independent through use of renewable resources and making of their own equipment; sponsors farms tours, workshops, slide presentations, library; newsletter and other publications.

NEW HAMPSHIRE

New Hampshire Solar Energy Association, c/o Chris Benz, Box 4382, Manchester, NH 03108. Local chapter of New England Solar Energy Association-American Section, International Solar Energy Society (see p. 78).

Northern New England Center for Appropriate Technology, 15 Garrison Avenue, Durham, NH 03824. Active in Maine, New Hampshire, and Vermont; promotes small-scale, low-energy, low-cost projects to increase self-sufficiency in energy, housing, and food production; provides information clearing house, workshops, conferences, media presentations, helps with government and business planning; runs project on home energy problems of elderly people; does training program for weatherization and energy conservation in homes; newsletter.

The Phoenix Nest, South Acworth, NH 03607. Nonprofit educational facility and residential cooperative; committed to developing skills and arts necessary for ecologically sound ways of living; runs both residential and nonresidential educational programs in appropriate technologies and the creative and performing arts (see text, p. 30); gives theatrical performances often combined with workshops in theater arts and alternative energy systems.

Society for the Protection of New Hampshire Forests, 5 South State Street, Concord, NH 03301. Educates and assists land owners and municipalities on wise use and management of forests and on wood as an energy source; provides library, literature, films, workshops, exhibits; newsletter and magazine; constructing new office building and demonstration center using alternative energy techniques.

TEA Foundation, Church Hill, Harrisville, NH 03450. Nonprofit affiliate formed to serve people unable to afford the research, design, and consulting services of Total Environmental Action, Inc. (see text, p. 22); programs include weatherizing low-income homes, training installers of solar equipment, developing sites for research, demonstration, and education projects related to energy conservation and the use of solar energy in buildings.

NEW JERSEY

Citizens Energy Council, Box 285, Allendale, NJ 07401. National organization, with representatives in 42 states conducts seminars; sponsors political action programs on developing energy from clean renewable resources, speakers, films; newsletter.

New Jersey Public Interest Research Group, 32 West Lafayette Street, Trenton, NJ 08608. Citizens' action group; does research, lobbying, educational programs, some concerned with using renewable energy sources.

NEW MEXICO

New Mexico Public Interest Research Group, 139 Harvard S.E., P.O. Box 4564, Albuquerque, NM 87106. Citizens' action group; gives educational programs on solar energy.

New Mexico Solar Energy Association, c/o Barbara Francis, P.O. Box 2004, Santa Fe, NM 87501. Regional chapter of American Section, International Solar Energy Society (see p. 78).

Solar Sustenance Team, Route 1, Box 107AA, Santa Fe, NM 87501. Small nonprofit educational organization; designs and constructs low-cost greenhouses; runs educational workshops, classes, construction workshops; has film and slide series.

Southwest Research and Information Center, P.O. Box 4524, Albuquerque, NM 87106. Citizens' resource group; maintains library on energy issues; magazine, *The Workbook*, and solar energy publications.

NEW YORK

American National Standards Institute, 1430 Broadway, New York, NY 10018. Private, nonprofit federation; coordinates the development of voluntary national standards, including solar energy standards; approves proposed standards as American National Standards; coordinates U.S. participation in voluntary, nongovernmental international standardization.

American Society of Heating, Refrigerating, and Air Conditioning Engineers, 345 East

47th Street, New York, NY 10017. Professional association; has technical committee on solar energy utilization; sponsors research and technical conferences on solar energy for heating and cooling; develops standards for testing solar-related equipment; publishes bulletins, journals, and handbooks that deal specifically with or contain information on use of solar energy for cooling and heating buildings.

Appropriate Technology Action Coalition, 156 Fifth Avenue, New York, NY 10010. Network of community housing and technical assistance groups involved with energy and appropriate technology in New York City; projects are operating or planned in neighborhoods of South Bronx, Lower East Side, Flatbush, Bedford-Stuyvesant, Corona, Williamsburg, and Harlem. (See "Energy Task Force," p. 23).

Center for Energy Policy and Research, New York Institute of Technology, Old Westbury, NY 11568. Under federal contract, operates an Information Center on energy conservation, alternative energy sources, and energy programs and policy; provides a clearinghouse service for the National Energy Extension Service; has videocassettes, literature.

Committee For Economic Development, 477 Madison Avenue, New York, NY 10022. Nonprofit organization; does research and disseminates information on practices of U.S. corporations in areas that vitally affect society, including development and use of energy resources; puts out reports and other publications.

Community Energy Network, 122 Anabel Taylor Hall, Ithaca, NY 14853. Community group; provides energy information through library, workshops, demonstrations at fairs, speakers, literature; helps with home installations.

Council on the Environment of New York City, 51 Chambers Street, Room 28, New York, NY 10007. Educational group for conservation and self-sufficiency; has information packet (free to residents of New York City) containing the following publications: *Community Energy Resources*, *Working with the Sun*, *Recycling Centers*, *Home Energy Conservation*, *Special Energy Conservation Projects*; lists groups active in all parts of NYC; has curriculum for grades 3 through 8, with home and school energy projects, lesson plans, work sheets and stencils.

Consumer Action Now, 355 Lexington Avenue, 16th Floor, New York, NY 10017. Solar advocacy group; organizes action to support passage of prosolar legislation, disseminates information on solar energy and conservation; has resource library, speakers; newsletter.

Eastern New York Solar Energy Society, P.O. Box 5181, Albany, NY 12205. Regional chapter of American Section, International Solar Energy Society (see p. 78).

Energy Task Force, 156 Fifth Avenue, New York, NY 10010. Community technical assistance organization; did first urban solar and wind energy systems to be used on a "sweat-equity" renovated tenement; now does installations of energy conserving and renewable energy systems in low-income neighborhoods, energy auditing, education and training programs.

Environment and Energy Resource Center of Western New York, Science and Engineering Library, Capen Hall, Amherst, NY 14260. Special library; provides energy information through library holdings, speakers, and newsletter.

INFORM, 25 Broad Street, New York, NY 10004. Research group; analyzes impact of U.S. corporations on employees, consumers, communities, and the environment; studies include energy ones, particularly reports on state-of-the-art of new energy technologies, which take in all solar-related energy sources.

Metropolitan New York Solar Energy Association, c/o Joseph Cuba, United Engineering Center, 345 East 47th Street, New York, NY 10017. Regional chapter of American Section, International Solar Energy Society (see p. 78).

National Center for Appropriate Technology Field Representative: Carlos Baez, 177 East 3rd Street, Store W, New York, NY 10009. See "Government" listings, p. 87.

People's Power Coalition, 196 Morton Avenue, Albany, NY 12208. State-wide coalition of grass-roots organizations; promotes alternative sources of energy through political action, training, consulting, speakers.

Solar Utilization in Northwest New York, c/o David Whitlock, P.O. Box 9501, Rochester, NY 14604. Local chapter of New England Solar Energy Association-American Section, International Solar Energy Society (see p. 78).

NORTH CAROLINA

North Carolina Public Interest Research Group, P.O. Box 2901, Durham, NC 27705. Student action group; does research and advocacy on several local issues, including energy conservation; sponsors library, speakers, conferences, media contacts; publishes research reports.

North Carolina Solar Energy Association, c/o Leon Neal, P.O. Box 12235, Research Triangle Park, NC 27709. Regional chapter of American Section, International Solar Energy Society (see p. 78).

OHIO

Ohio Solar Energy Association, 389 East 271st Street, Euclid, OH 44132. Regional chapter of American Section, International Solar Energy Society (see p. 78). With other local chapters forming, these are the current ones:

Miami Valley Alternative Energy Association, P.O. Box 3087, Overlook Station, Dayton, OH 45431
Northeast Ohio Alternative Energy Association, 389 East 271st Street, Euclid, OH 44132
Southwest Ohio Alternative Energy Association, 129 Laura Lane, Cincinnati, Ohio 45233

Sandusky Center for Appropriate Technology, 4620 Hayes Avenue, Sandusky, OH 44870. Community appropriate technology group; has test and demonstrations project in collector system for hot water heating.

OKLAHOMA

Oklahoma Solar Energy Association, c/o Dr. Bruce V. Ketcham, Solar Energy Laboratory, University of Tulsa, Tulsa, OK 74104. Regional chapter of American Section, International Solar Energy Society (see p. 78).

OREGON

Cascadian Regional Library (CAREL), 1 West Fifth Avenue, Box 1492, Eugene, OR 97440. Includes all nonprofit and community action organizations in Pacific Northwest; acts as information network in 20 categories, one of which is energy; organizes conferences; collects and disseminates information; publishes *Cascade: Journal of the Northwest*, which reports activities of Northwest organizations.

Cerro Gordo Ranch, Dorena Lake, P.O. Box 569, Cottage Grove, OR 97424. Cooperative community of over 100 members; building toward self-sufficient community of 2,500 people on 1,200 acres near Eugene, OR, using appropriate technology; has plans, library, newsletter and magazine describing this and other experiments.

Lane County Housing & Energy Division, 170 East 11th Street, Eugene, OR 97401. Local government agency carrying out energy management program for County; provides training and demonstration projects in weatherization and solar technologies; maintains library on solar and other appropriate technologies.

Portland Community Design Center, 723 S.E. Grand Avenue, Portland, OR 97214. Service agency; provides low-cost architectural and environmental design assistance to low-income individuals and nonprofit groups in metropolitan Portland area; assists with renovation, preservation, and remodeling of existing structures, neighborhood planning, design of parks, playgrounds, structures using energy conservation principles; runs depot for reusable building materials.

Rain: Journal of Appropriate Technology, 2270 N.W. Irving, Portland, OR 97210. Citizens' research and information group; publishes a variety of literature, including *Rain*, on appropriate technology and self-reliance; consults on energy in Montana, Idaho, Oregon, Washington, and California; participated in formation of National Center for Appropriate Technology; has wind, solar, wood stove, and economic policy information.

Solar Energy Center, University of Oregon, Eugene, OR 97403. University group; promotes use of solar energy and conservation through academic programs and research; runs biweekly public seminars; has speakers, workshops, demonstration projects, publications, information and resource center for the community on solar energy and conservation.

PENNSYLVANIA

American Society for Testing and Materials, 1916 Race Street, Philadelphia, PA 19103. Nonprofit corporation with over 28,000 members internationally; develops standards on characteristics and performance of materials, products, systems, and services; Committee on Solar Energy Conversion works to develop standards for solar energy applications, including active and passive space heating and cooling, hot water and swimming pool heating, process heating, thermal conversion power generation, photovoltaics, ocean thermal, wind energy, and biomass.

Mid-Atlantic Solar Energy Association, c/o Robert T. Bennett, Department of Architecture, University of Pennsylvania, Philadelphia, PA 19104. Regional chapter of American Section, International Solar Energy Society (see p. 78).

National Center for Appropriate Technology Field Representative: Beth Hopkins, 131 Howard Avenue, Lancaster, PA 17602. See "Government" listings, p. 87.

Society for Energy Alternatives in the Urban-Rural Environment, 335 Saunders Avenue, Philadelphia, PA 19104. Citizens' group acts as resource center for alternative energy systems; holds weekly open meetings; retrofitted a town house for solar systems; provides library, educational demonstrations for fairs and meetings.

RHODE ISLAND

Rhode Island Solar Energy Association, c/o George Jennings, P.O. Box 212, Providence, RI 02901. Local chapter of New England Solar Energy Association-American Section, International Solar Energy Society (see p. 78).

TENNESSEE

Tennessee Environmental Council, P.O. Box 1422, Nashville, TN 37202. Group of citizens and organizations; provides support for solar energy and energy conservation and information to homeowners and other citizens through conferences, workshops and technical advice; newsletter.

TEXAS

Citizens' Environmental Coalition Educational Fund, Inc., 1 Main Plaza S-1006, Houston, TX 77002. Represents over 40 community groups; serves as clearinghouse for information and activities on energy conservation, renewable energy sources, and other environmental topics; has library, literature, workshops, exhibits, speakers; newsletter.

The Center for Maximum Potential Building Systems, 8604 Webberville Road, Austin, TX 78724. Small nonprofit research and educational facility; works on energy and resource conservation and on serving as regional appropriate technology laboratory; engages in building projects using local materials and renewable resources (such as solar farm, passive solar adobe); does formal presentations, social and political planning, projects for low-income people; has slides and publications.

The Energy Laboratory, University of Houston, Houston, TX 77004. University group; disseminates information on all aspects of energy, including conservation and solar; provides workshops, seminars, conferences, technical and economic studies, demonstration projects, speakers; has Department of Energy contract to search out energy education materials for grades K through 12; newsletter.

International Solar Energy Society—American Section, American Technological University, P.O. Box 1416, Killeen, TX 76541. The most comprehensive association in the field, founded in 1954, now has 8,000 members in 55 countries; provides a forum for scientists, engineers, architects, contractors, manufacturers, investors, and legislators, as well as for home owners and other individuals interested in the advancement of solar energy; keeps up on the latest technical developments, legislation, and educational programs and disseminates this information through publications and major conferences and exhibits; *Solar Age*, official magazine. Regional and local chapters of the American Section (an excellent source of information about solar activity in their locations and welcoming of participation by interested individuals) are listed by individual states; sponsor meetings, conferences, and workshops; have libraries and literature available; often publish newsletters.

National Center for Appropriate Technology Field Representative: Dee Simpson, 2521 Rogers, Fort Worth, TX 76109. See "Government" listings, p. 87.

Solar Engineering Magazine, 8435 North Stemmons Freeway, Suite 880, Dallas, TX 75247. Official publication of the Solar Energy Industries Association (see p. 71); sponsors technical seminars, workshops, demonstration projects, exhibits; has library, speakers, slides, and videocassettes on solar energy.

Texas Solar Energy Society, c/o Peter E. Jenkins, Texas A & M University, College Station, TX 77843. Regional chapter of American Section, International Solar Energy Society (see p. 78).

VERMONT

Energy Responsibility, Box 411, Castleton, VT 05735. Community group; uses and develops use in surrounding towns of alternative energy forms and conservation practices; sponsors forums, workshops, trips to observe alternative energy use; chooses monthly energy theme appropriate to particular Vermont climate and season.

New England Regional Energy Project, P.O. Box 514, Burlington, VT 05401. Federally funded center; gives technical assistance and legal aid for energy issues; has library, information clearinghouse, workshops; newsletter.

New England Solar Energy Association, P.O. Box 541, Brattleboro, VT 05301. Regional chapter of American Section, International Solar Energy Society (see p. 78).

Solar Association of Vermont, Box 732, Montpelier, VT 05602. Local chapter of New England Solar Energy Association-American Section, International Solar Energy Society (see p. 78).

Southeastern Vermont Community Action, Box 396, Bellows Falls, VT 05101. Federally funded community agency; trains and hires unemployed to make stoves; has demonstration projects, speakers, workshops, exhibits on wood heating and some on solar; newsletter.

Conservation Society of Southern Vermont, Townshend, VT 05353. Environmental group; takes leadership in exploring use of renewable resources and development of regional self-sufficiency; provides resource center, conferences, community discussions, studies, and reports.

VIRGINIA

American Gas Association, 1515 Wilson Boulevard, Arlington, VA 22209. Solar Energy Committee sponsors seminars; serves as focus for several companies operating solar installations; distributes booklet describing solar utilization by the industry.

Sheet Metal and Air Conditioning Contractors National Association, Inc., 8224 Old Court House Road, Vienna, VA 22180. Trade association; sets standards, specifications for equipment; works on building codes and installer certification; refers inquiries to local contractors experienced in solar installations; distributes home study course on fundamentals of solar heating and cooling and other publications; puts on conferences.

Virginia Solar Energy Association, c/o John G. Lewis, Jr., P.O. Box 12442, Richmond, VA 23241. Regional chapter of American Section, International Solar Energy Society (see p. 78).

WASHINGTON

Ecotope Group, Inc., 2332 East Madison, Seattle, WA 98112. Does research and education in Pacific Northwest applications of appropriate technology, wind, solar; built methane generator, aquaculture-greenhouse systems; gives workshops on methane, solar hot water; does solar design and feasibility studies; provides library, speakers, consulting.

National Center for Appropriate Technology Field Representative: Birny Birnbaum, 1505 10th Avenue, Seattle, WA 98122. See "Government" listings, p. 87.

Soap Lake Solar Community Development Committee, Chamber of Commerce, P.O. Box 433, Soap Lake, WA 98851. Community research and solar development association; retrofitted local Chamber of Commerce building and high school; plans community conversion to solar power; makes available annual solar conference, speakers, workshops, information center.

WISCONSIN

Community Builders Coop, 2796 North Nicolet Drive, Green Bay, WI 54301. Small group; consults, designs, and builds solar, with emphasis on passive applications and owner-builder approaches; provides speakers, exhibitors, educators, and cooperators.

Wisconsin Association for Environmental Education, Inc., 125 West Kohlet Street, Sun Prairie, WI 53590. Volunteer organization; promotes energy education through conferences, workshops, exhibits; newsletter.

WYOMING

State Energy Conservation Office, 320 West 25th Street, Cheyenne, WY 82002. Distributing funds (not exceeding $10,000) to 30 communities for their local energy conservation efforts and use of alternative, renewable sources of energy; contact office for names of communities with groups starting.

Canada

Canadian Renewable Energy News, P.O. Box 4869 Station E, Ottawa, Ontario K1S 5B4. Has information about activities in energy conservation and use of renewable resources throughout Canada.

The Institute of Man and Resources, 50 Water Street, P.O. Box 2008, Charlottetown, Prince Edward Island C1A 1A4. Develops, tests, and assesses systems for transportation, food, and shelter based on wood, solar, wind, water, biomass, and other renewable resources; manages research and demonstration projects at New Alchemy Ark, a prototype of shelter design and biologically balanced indoor and outdoor food production (see text, p. 21); helps develop manufacturing and market potentials; provides assistance and information to individuals, groups, communities, and the government; provides literature, workshops, conferences, resource library.

Solar Energy Society of Canada, Inc., 608-870 Cambridge Street, Winnipeg, Manitoba R3M 3H5. Founded in 1974 to foster solar technologies, SESCI has grown to 3,000 members and in 1979 became a member of the International Solar Society; runs a large annual solar conference in various Canadian centers (with the technical and policy papers presented being published and serving as one of the most comprehensive sources of information on the status of solar energy development in Canada and internationally); maintains an information service; newsletter, "Sol." Chapters have been formed at the local level to provide talks, seminars, local newsletters, and other activities that promote solar energy and educate the public.

Alberta
SESCI Northern Alberta Chapter, 1104 81st Avenue, Edmonton, Alberta T6G 0S3
SESCI Peace Region Chapter, Box 204, Grande Prairie, Alberta T8V 3A4

British Columbia
SESCI British Columbia Chapter, c/o Richard Kadulski, Box 34308, 2405 Pine Street, Vancouver, British Columbia V6J 4P3
SESCI Vancouver Island Chapter, 235 Wilson Street, Victoria, British Columbia

Manitoba
SESCI Manitoba Chapter, 608-870 Cambridge Street, Winnipeg, Manitoba R3M 3H5

New Brunswick
SESCI New Brunswick Chapter, 569 Canterbury Drive, Fredericton, New Brunswick E3B 4M6

Ontario
SESCI Barrie Chapter, 73 St. Vincent Street, Barrie, Ontario L4M 3Y5
SESCI Kawartha Chapter, c/o Judy Shaw, 250 Aberdeen Avenue, Peterborough, Ontario K9H 2W9
SESCI Kent Chapter, Apartment 807-150 Mary Street, Chatham, Ontario N7L 4V2
SESCI Kingston Chapter, P.O. Box 153, Station "A," Kingston, Ontario K7M 6R1
SESCI Mississauga-Oakville Chapter, 2222 Belfast Crescent, Mississauga, Ontario L5K 1N9
SESCI Ottawa Chapter, c/o Algonquin College Physics Department, Lees Avenue, Ottawa, Ontario K1S 6R1
SESCI Sarnia Chapter, c/o W. Himmelman, 1034 Lombardy Drive, Sarnia, Ontario N7S 2E2
SESCI Thunder Bay Chapter, c/o E. J. Tymura, Hilldale Road, Thunder Bay, Ontario P7B 5N1
SESCI Toronto Chapter, P.O. Box 396, Station "D," Toronto, Ontario M6P 3L9
SESCI Windsor-Essex Chapter, Chemical Engineering Department, University of Windsor, Windsor, Ontario N9B 3P4

Prince Edward Island
SESCI Prince Edward Island Chapter, P.O. Box 1041, Charlottetown, Prince Edward Island C1A 7M4

Quebec
SESCI Section Quebecoise, Department Genie Mecanique, Ecole Polytechnique, C.P. 6079, Succ. A., Montreal, Quebec H3C 3A7

Saskatchewan
SESCI Northern Saskatchewan Chapter, Department Mechanical Engineering, University of Saskatchewan, Saskatoon, Saskatchewan S7N 0W0
SESCI Regina Chapter, P.O. Box 3959, Regina, Saskatchewan S4P 3R9
SESCI Swift Current Chapter, 416 Hayes Drive, Swift Current, Saskatchewan S9H 4H9

Government

Federal and state government offices are likely sources of jobs. They also have information about solar enterprises, both government-related and independent, in particular geographic locations and fields of expertise.

STATE ENERGY OFFICES

The state energy offices are listed below. The activities and types of jobs found in these offices are summarized in the text on page 27.

Alabama Energy Management Board
State Capitol
Montgomery, AL 36130
(205) 832-5010

Alaska State Energy Office
Mackay Building, 7th Floor
338 Denali Street
Anchorage, AK 99501
(907) 272-0527

Arizona Solar Energy Research Commission
1700 West Washington, Room 502
Phoenix, AZ 85007
(602) 271-3682

Arkansas State Energy Office
Commerce Department
960 Plaza West Building
Little Rock, AR 72205
(501) 371-1379

California Energy Resources
Conservation and Development Commission
1111 Howe Avenue
Sacramento, CA 95825
(916) 322-3690

Colorado Office of Energy Conservation
1410 Grant Street, Suite B104
Denver, CO 80203
(303) 839-2507

Connecticut Department of Planning and
Energy Policy
20 Grand Street
Hartford, CT 06115
(203) 566-2800

Delaware Governor's Energy Office
Townsend Building
P.O. Box 1401
Dover, DE 19901
(302) 678-5644

Florida State Energy Office
301 Bryant Building
Tallahassee, FL 32304
(904) 488-6764

Georgia Office of Energy Resources
270 Washington Street S.W., Suite 615
Atlanta, GA 30334
(404) 656-5176

Hawaii Department of Planning & Economic
Development
P.O. Box 2359
Honolulu, HI 96804
(808) 548-3033

Idaho Office of Energy
State Capitol
Boise, ID 83720
(208) 384-3258

Illinois Division of Energy
Department of Business & Economic
Development
222 South College Avenue
Springfield, IL 62706
(217) 782-7500

Indiana Energy Office
Consolidated Building, 7th floor
115 North Pennsylvania
Indianapolis, IN 46204
(317) 633-6753

Iowa Energy Policy Council
215 East 7th Street
Des Moines, IA 50319
(515) 247-4420

Kansas Energy Office
503 Kansas Avenue, Room 241
Topeka, KS 66603
(913) 296-2496

Kentucky Department of Energy
Capitol Plaza Tower
Frankfort, KY 40601
(502) 564-7416

Energy Division of Louisiana
Research & Development
Division of Natural Resources
P.O. Box 44156
Baton Rouge, LA 70804
(504) 389-2771

Maine Office of Energy Resources
55 Capitol Street
Augusta, ME 04333
(207) 289-2195

Maryland Energy Policy Office
301 West Preston Street, Suite 1302
Baltimore, MD 21202
(301) 383-6810

Massachusetts Energy Policy Office
73 Tremont Street, Room 700
Boston, MA 02108
(617) 727-4732

Michigan Energy Administration
Law Building, 4th Floor
Lansing, MI 48913
(517) 374-9090

Minnesota Energy Agency
American Center Building, 9th Floor
150 East Kellogg Boulevard
St. Paul, MN 55101
(612) 298-5120

Mississippi Energy Office
1307 Woolfolk State Office Building
Jackson, MS 39202
(601) 354-7406

Missouri Energy Agency
P.O. Box 176
1014 Madison Street
Jefferson City, MO 65101
(314) 751-4000

Montana Energy Office
Capitol Station
Helena, MT 59601
(406) 449-3940

Nebraska State Energy Office
301 Centennial Mall
Lincoln, NE 68509
(402) 471-2867

Nevada Department of Energy
1050 East William, Suite 405
Carson City, NV 89701
(702) 885-5157

New Hampshire Energy Office
26 Pleasant Street, 3rd Floor
Concord, NH 03301
(603) 271-2711

New Jersey State Energy Office
101 Commerce Street
Newark, NJ 07102
(201) 684-3290

New Mexico Energy Resources Board
P.O. Box 2770
Santa Fe, NM 87503
(505) 827-2472

New York Energy Office
Agency Building 2
Empire State Plaza
Albany, NY 12223
(518) 474-8313
toll free hotline for energy information:
(800) 342-3722

North Carolina Energy Policy Council
P.O. Box 25249
215 East Lane
Raleigh, NC 27611
(919) 733-2230

North Dakota Energy Office
1533 North 12th Street
Bismarck, ND 58501
(701) 224-2250

Ohio Department of Energy
30 East Broad Street, 34th Floor
Columbus, OH 43215
(614) 466-6797

Oklahoma Department of Energy
4400 North Lincoln Boulevard, Suite 251
Oklahoma City, OK 73105
(405) 521-3941

Oregon Department of Energy, Labor and
Industries Building, Room 111
Salem, OR 97310
(503) 378-4128

Pennsylvania Energy Office
State Office Building, 12th Floor
500 North 3rd Street
Harrisburg, PA 17101
(717) 787-9749

Rhode Island State Energy Office
80 Dean Street
Providence, RI 02903
(401) 277-3370

South Carolina Energy Management Office
1205 Pendleton Street
Columbia, SC 29201
(803) 758-2050

South Dakota Office of Energy Policy
Anderson Building
Pierre, SD 57501
(605) 224-3603

Tennessee Energy Office
250 Capitol Hill Building
Nashville, TN 37219
(615) 741-2994

Texas Governor's Energy Office
7703 North Lamar, Room 502
Austin, TX 78752
(512) 475-5491

Utah Energy Office
455 East 4th Street South, Suite 303
Salt Lake City, UT 84111
(801) 533-5424

Vermont State Energy Office
Pavilion Building, 5th Floor
Montpelier, VT 05602
(802) 828-2768

Virginia Energy Office
823 East Main Street
Richmond, VA 23228
(804) 786-8451

Washington State Energy Office
400 East Union, 1st Floor
Olympia, WA 98504
(206) 753-2417

West Virginia Fuel and Energy Office
1262 Greenbrier Street
Charleston, WV 25311
(304) 348-8860

Wisconsin Office of State Planning and Energy
P.O. Box 511
Madison, WI 53701
(608) 266-8234

Wyoming Energy Conservation Office
320 West 25th Street
Cheyenne, WY 82002
(307) 777-7131

FEDERAL

The federal government is a large employer in all fields, and solar energy is no exception. People work in Washington, D.C., in laboratories and agencies located outside the capitol, and in state branches of federal departments. Virtually all occupations are represented in federal solar employment. Many of the jobs require an interest and ability in administrative work.

In this section, the first listings are of five special solar programs sponsored by the federal government expressly for the development of renewable resources and energy conservation. The second listings, federal departments and offices, start with the U.S. Department of Energy, which has the majority of the solar projects, and proceeds through many other solar-related agencies. Third, this section gives the names of congressional committees that have some influence over solar development and identifies the members of the Congressional Solar Coalition.

Special Solar Programs

Regional Solar Development Centers

Four centers have been funded by the U.S. Department of Energy to foster the widespread use of solar energy by means of regionally diversified efforts. Each center's mission is to match the solar energy resources of the states in its area to the energy needs of the region. Centers work with industrial, architectural, construction, agricultural, academic, financial, legal, real estate, and insurance communities, as well as with trade unions and utilities. It is expected that the efforts of these centers will not only reduce the regions' dependence on oil, but also will stimulate the economy and create new jobs in solar industries. (See text, p. 41.)

Mid-American Solar Energy Complex, 1256 Trapp Road, Eagan, MN 55121

Illinois	Kansas	Missouri	Ohio
Indiana	Michigan	Nebraska	South Dakota
Iowa	Minnesota	North Dakota	Wisconsin

Northeast Solar Energy Center, 70 Memorial Drive, Cambridge, MA 02142

Connecticut	New Hampshire	Pennsylvania
Maine	New Jersey	Rhode Island
Massachusetts	New York	Vermont

Southern Solar Energy Center, Exchange Place, Suite 1250, 2300 Peachford Road, Atlanta, GA 30338

Alabama	Florida	Mississippi	Tennessee
Arkansas	Georgia	North Carolina	Texas
Delaware	Kentucky	Oklahoma	Virgin Islands
District of Columbia	Maryland	South Carolina	Virginia

Western SUN, Portland, OR

Alaska	Hawaii	Nevada	Utah
Arizona	Idaho	New Mexico	Washington
California	Montana	Oregon	Wyoming
Colorado			

Energy Extension Service

Department of Energy, Washington, DC 20585

The Department of Energy has set up energy extension services in 10 states as pilot programs, and will cover other states in the near future. The services give information and assistance on energy conservation and the use of renewable energy resources to individuals, groups, businesses, and government agencies, and help coordinate other energy programs in the states where they are located.

The following list identifies the on-going pilot programs and provides the names of the people to contact in states where Energy Extension Services are in the process of being developed.

Pilot Program
John Hoyle, Director
Alabama Energy Extension Service
Auburn University
Auburn, AL 36830

John Hale
MacKay Building, 7th floor
338 Denali Street
Anchorage, AK 99501

Donald E. Osborn
Arizona Solar Energy Research Commission
1700 West Washington
Phoenix, AZ 85007

Donald Bailey
Arkansas Energy Office
960 Plaza West
Little Rock, AR 72205

Allen Lind
Office of Appropriate Technology
1530 Tenth Street
Sacramento, CA 95814

Robert Brown
Colorado Office of Energy Conservation
1600 Downing Street
Denver, CO 80218

Pilot Program
Bradford S. Chase, Director
Connecticut Energy Extension Service
Office of Policy and Management
80 Washington Street
Hartford, CT 06115

Clarence Stukes
Office of Planning and Development
1329 E Street, NW, Suite 600
Washington, DC 20004

Gail A. Lanouette
Governor's Office
Townsend Building
P.O. Box 1401
Dover, DE 19901

Jim Jackson
State Energy Office
301 Bryant Building
Tallahassee, FL 32304

Shannon St. John
Office of Energy Resources, Room 615
270 Washington Street, SW
Atlanta, GA 30334

Ed Greaney
State Energy Office
1164 Bishop Street, Suite 1515
Honolulu, HI 96813

L. Kirk Hall
Idaho Office of Energy
State House
Boise, ID 83720

Webster Bay
State of Illinois Institute of Natural Resources
325 W. Adams Street, Room 300
Springfield, IL 62706

Robert J. Hedding
Indiana Energy Group
Indiana Department of Commerce
Consolidated Building, 7th Floor
115 N. Pennsylvania Street
Indianapolis, IN 46204

H. K. Baker
110 Marston Hall
Iowa State University
Ames, IA 50011

John Rowe
Kansas Energy Office
503 Kansas Avenue, Room 241
Topeka, KS 66603

John Stapleton
Bureau for Energy, Production, and Distribution
Capital Plaza Tower
Frankfort, KY 40601

Vincent DiCara
Office of Energy Resources
55 Capitol Street
Augusta, ME 04330

Felicity Evans
Energy Policy Office, Suite 1302
301 West Preston Street
Baltimore, MD 21201

Sandra Turner
Office of Energy Resources
73 Tremont Street
Boston, MA 02108

Pilot Program
Dennis Sykes, Director
Michigan Energy Extension Service
6520 Mercantile Way, Suite 1
Lansing, MI 48910

Tom Stern
Minnesota Energy Agency
American Center Building
150 East Kellogg Boulevard
St. Paul, MN 55101

Dr. William W. Linder
Cooperative Extension Service
Mississippi State University
Box 5406
Mississippi State, MS 39762

Robert Buell
Missouri Energy Program
Department of Natural Resources
P.O. Box 176
Jefferson City, MO 65102

Paul Cartwright
Energy Division
Department of Natural Resources and Conservation
32 S. Ewing
Helena, MT 59601

Donald M. Edwards
Energy Research and Development Center
W181 Nebraska Hall
University of Nebraska-Lincoln
Lincoln, NE 68588

Noel A. Clark
Nevada Department of Energy
1050 East William, Suite 405
Capitol Complex
Carson City, NV 89710

William R. Humm
Governor's Council on Energy
2½ Beacon Street
Concord, NH 03301

Ira H. Dorfman
Office of Conservation
New Jersey Department of Energy
101 Commerce Street
Newark, NJ 07102

Pilot Program
George H. Anderson, Director
New Mexico Energy Extension Service
P.O. Box 00
Santa Fe, NM 87501

Sandy Schuman
New York State Energy Office
2 Rockefeller Plaza
Albany, NY 12223

Brian M. Flattery
Energy Division
North Carolina Department of Commerce
430 N. Salisbury Street
Raleigh, NC 27611

Bruce Westerberg
Office of Energy Management and Conservation
1533 North 12th Street
Bismarck, ND 58501

Thomas P. Ryan
Ohio Department of Energy
30 East Broad Street
Columbus, OH 43215

Patty Miller
Oklahoma Department of Energy
4400 North Lincoln Boulevard, Suite 251
Oklahoma City, OK 73105

Owen D. Osborne
Oregon State University
Corvallis, OR 97331

Pilot Program
John Hafer, Director
Pennsylvania Energy Extension Service
Governor's Energy Council
1625 North Front Street
Harrisburg, PA 17102

Carmen Martha Farre
Puerto Rico Office of Energy
Minillas Governmental Center
P.O. Box 41089, Minillas Station
Santurce, PR 00940

Vin Graziano
Governor's Energy Office
80 Dean Street
Providence, RI 02903

Steve Gomez
Office of Energy Policy
Capital Lake Plaza
Pierre, SD 57501

Pilot Program
Douglas Bennett, Jr., Director
Tennessee Energy Extension Service
226 Capitol Boulevard Building, Suite 615
Nashville, TN 37319

Pilot Program
Dr. Stephen Riter
Texas Energy Extension Service
Texas A&M University
College Station, TX 77849

Douglas E. Thompson
Utah Energy Office
231 East 400 South, Suite 101
Salt Lake City, UT 84111

Bruce Haskell
Pavilion Office Building
Montpelier, VT 05602

J. Boyd Spencer
State Office of Energy and Energy Services
310 Turner Road
Richmond, VA 23225

Dr. Michael J. Canoy
Energy Study
College of the Virgin Islands
St. Thomas
U.S. Virgin Islands 00801

Pilot Program
J. Orville Young, Director
Washington Energy Extension Service
Cooperative Extension Service
AG-Phase II
Washington State University
Pullman, WA 99164

Ron Potesta
West Virginia Fuel and Energy Office
1262½ Greenbriar Street
Charleston, WV 25311

Pilot Program
William Bernhagen, Director
Wisconsin Energy Extension Service
University of Wisconsin Extension
432 North Lake Street, Room 435
Madison, WI 53706

Pilot Program
Sam D. Hakes, Director
Wyoming Energy Extension Service
University Station, Engineering Bldg.
Box 3295
Laramie, WY 82071

Community Services Administration (CSA)
Energy Conservation Program
200 Nineteenth Street NW, Washington, DC 20506

CSA is the main agency supporting energy projects for low-income groups in both urban and rural settings. It provides funds for energy staff in 900 local Community Action Programs (CAPs) in 50 states. Programs include education, research and development, job training, emergency help with fuel bills, and application of appropriate technologies such as weatherizing, domestic solar hot water, wind power, and solar greenhouses. Local CAPs welcome volunteers who want to get involved in energy conservation and renewable energy and they engage with the community in their projects.

National Center for Appropriate Technology (NCAT)
P.O. Box 3838, Butte, MT 59701

NCAT was established by the CSA to apply small-scale, low cost technologies to the food and energy needs of low-income people. It operates national information sharing networks, conferences, and workshops; funds regional newsletters and community projects; provides research and technical assistance in several areas of alternative energy use; maintains a resource center and library, and issues technical reports and pamphlets.

Contact between the NCAT office in Butte and all regions of the United States is provided by the following field representatives, who also have information on other solar-related organizations in their regions:

Region I — New England
Robert Shortreed
84 Westford Road
Stafford Springs, CT 06076
(203) 429-0160

Region II — Northeast
Carlos Baez
177 East 3rd Street (Store W)
New York, NY 10009
(212) 475-4840

Region III — Appalachia
Beth Hopkins
131 Howard Avenue
Lancaster, PA 17602
(717) 299-1845

Region IV — Southeast
Tyrone Brooks
1315 Beecher Street
Southwest Atlanta, GA 30310
(404) 522-1420
753-3361

Region V — Midwest
Jim Laukes
105 Wolpers Road
Park Forest, IL 60466
(312) 481-6168

Region VI — Southwest
Dee Simpson
2521 Rogers
Fort Worth, TX 76109
(817) 923-9445

Region VII — Plains
Council Smith
The National Center for Appropriate Technology
438 N. Skinker
St. Louis, MO 63130
(314) 721-8487

Region VIII — Rocky Mountains
Doug Baston
Rte. 2, Box 725C
Coal Creek Canyon
Golden, CO 80401
(303) 642-7584

Region IX — West Coast
David Colfax
The Mountain School
P.O. Box 246
Boonville, CA 95415
(707) 895-3755

Region X — Northwest
Birny Birnbaum
1505 10th Avenue
Seattle, WA 98122
(206) 324-5005

National Solar Heating and Cooling Information Center
P.O. Box 1607, Rockville, MD 20850

Operated for the U.S. Department of Housing and Urban Development and the Department of Energy as part of the Franklin Institute, the center supplies technical and nontechnical information to consumers, builders, suppliers, and indeed anyone who needs it. The center answers inquiries by phone, or sends out sheets of information, bibliographies, and answers to specific questions. Toll free number: (800) 523-2929; (800) 462-4983 in Pennsylvania only.

Solar Energy Research Institute
1536 Cole Boulevard, Golden, CO 80401

Funded by the Department of Energy, the Institute conducts research in basic solar technologies and in methods of implementing these technologies, with the goal of creating major alternative sources of energy for the United States. (See text, p. 48).

Departments and Offices

Federal departments have several levels of administration, starting with the secretary of the department, a cabinet post, and continuing through assistant secretaries and deputy assistant secretaries. The listings in this section omit some of the superstructure in order to focus on the offices that are relevant to solar energy.*

U.S. Department of Energy
Washington, DC 20585, (202) 376-4000

This department is responsible for the development of all sources of energy, including nuclear and fossil fuels. It has a number of programs promoting solar energy and energy conservation, in addition to the ones listed among "Special Solar Programs" above.

Office of Conservation and Solar Applications
(202) 376-4934, Acting Assistant Secretary: Dr. Maxine Savitz.

Office of Commercialization
Director: J. L. Barrow, Jr.

Office of Buildings and Community systems
Director: J. P. Millhone.

---

* The *United States Government Manual*, the official handbook of the federal government, contains detailed information on the structure of the various departments and agencies. It is available for $6.50 from Superintendent of Documents, U.S. Government Printing Office, Washington, DC 20402.

Office of Industrial Programs
Director: D. G. Harvey.

Office of Transportation Programs
Acting Director: P. J. Brown.

Office of Solar Applications
Director: F. H. Morse.

Office of State and Local Programs
Director: F. M. Stewart.

Office of Small Scale Technology
Director: W. Otis.

Office of Energy Technology
(202) 252-6842, Assistant Secretary: Dr. John M. Deutch.

Office of Solar, Geothermal, Electric, and Storage Systems
(202) 376-9190, Director: Bennett Miller.

Division of Central Solar Technology — solar thermal and ocean systems. Director: Howard Coleman.

Division of Distributed Solar Technology — biomass, photovoltaic, and wind energy systems. Director: Robert San Martin.

Division of Geothermal Energy — geothermal and hydrothermal energy. Director: Bennie DiBona.

Division of Electric Energy Systems — power delivery and power supply integration. Director: F. F. Parry.

Division of Energy Storage Systems — electrochemical, thermal, and mechanical systems. Director: George Perdirtz.

Division of Planning and Technical Transfer — strategy and analysis, environmental concerns. Acting Director: Leslie Levine.

Division of Program Resource Management — administrative operations, financial resources. Director: Richard Benson.

Information available from Department of Energy, Office of Public Affairs:

"Solar Weekly Announcements," covering solar technology program developments within the Department of Energy.

"Solar Energy Research and Development Report," monthly.

"Solar Energy Update," abstract journal, monthly.

The Department of Energy maintains ten regional offices:

Region I — New England
U.S. Department of Energy
150 Causeway Street, Room 700
Boston, MA 02114
(617) 223-3703

Region II — Northeast
U.S. Department of Energy
26 Federal Plaza, Room 3200
New York, NY 10007
(212) 264-0119

Region III — Appalachia
U.S. Department of Energy
1421 Cherry Street, Room 1001
Philadelphia, PA 19102
(215) 597-9066

Region IV — Southeast
U.S. Department of Energy
1655 Peachtree Street, N.E.
Atlanta, GA 30309
(404) 881-2062

Region V — Midwest
U.S. Department of Energy
Insurance Exchange Building
175 W. Jackson Boulevard
Chicago, IL 60604
(312) 353-5760

Region VI — Southwest
U.S. Department of Energy
P.O. Box 35228
Dallas, TX 75235
(214) 749-7714

Region VII — Plains
U.S. Department of Energy
324 East 11th Street
Kansas City, MO 64106
(816) 374-2064

Region VIII — Rocky Mountains
U.S. Department of Energy
P.O. Box 26249, Bellmawr Branch
Lakewood, CO 80226
(303) 234-3378

Region IX — West Coast
U.S. Department of Energy
215 Fremont Street, 6th Floor
San Francisco, CA 94105
(415) 556-7216

Region X — Northwest
U.S. Department of Energy
1992 Federal Building
915 Second Avenue
Seattle, WA 98174
(206) 442-7285

U.S. Department of Housing and Urban Development (HUD)
451 Seventh Street, NW, Washington, DC 20410, (202) 755-5111

Office of Policy Development and Research
(202) 755-5600, Assistant Secretary: Ms. Donna E. Shalala.

Office of Research and Demonstration
(202) 755-5561, Deputy Asst. Secretary: Raymond Struyk.

Office of Energy, Building Technology, and Standards,
(202) 755-6900.

Through this office, HUD funds and monitors many demonstration projects using solar heating and cooling in buildings. It also implements energy conservation standards and codes.

Building Technology Research Staff — Program Manager: Orville Lee

Energy Standards Research Staff — Program Manager: vacant

Solar Heating and Cooling Research Staff — Program Manager: David Moore

U.S. Department of Commerce
14th Street and Constitution Avenue NW,
Washington, DC 20230, (202) 377-2000

National Bureau of Standards
Washington, DC 20234, (301) 921-1000.

The bureau initiates and maintains guidelines and standards to protect the public and the environment.

Solar Technology — Group Leader: Robert D. Dikkers
Thermal Solar — Group Leader: Dr. James Hill

National Oceanic and Atmospheric Administration
6010 Executive Boulevard, Rockville, MD 20852, (301) 655-4000.

Environmental Research Laboratories, Air Resources Laboratory, 310 Marine Avenue, Boulder, CO 80302, (303) 499-1000, Director: Dr. Lester Machta.

Maintains nationwide network of stations to record radiation.

Environmental Research Laboratories,
Geophysical Monitoring for Climatic Change Program, GRAMAX Building, Silver Spring, MD 20910, (301) 427-7645, Director: Dr. Kirby Hanson.

Monitors radiation at remote locations and calibrates instruments for solar radiation measurement.

Environmental Data and Information Service, National Climatic Center, Federal Building, Asheville, NC 28801, (704) 258-2850, Director: Daniel B. Mitchell.

Collects archives and publishes solar radiation data monthly for national weather service reports; puts out solar radiation information in written reports and on computer tapes.

U.S. Department of Agriculture

Fourteenth Street and Independence Avenue SW,
Washington, DC 20250, (202) 447-3987

Energy Program

Beltsville, MD, (301) 344-2740, Program Manager: Dr. Landry Altman.

Has 50 research and development projects (funded by Department of Energy) covering use of solar energy for greenhouses, rural residences, crop and grain drying, livestock shelter, and food processing.

U.S. Department of Defense

The Pentagon, Washington, DC 20301, (202) 545-6700

Office of Installations and Housing

Deputy Assistant Secretary: Perry J. Fliakas.

Office of Engineering Standards and Design
(202) 695-7804, Director: Mortimer M. Marshall.

Responsible for energy conservation and installation of solar equipment on military-owned buildings.

Agencies (not part of departmental structure)

Federal Home Loan Bank Board

1700 G Street NW, Washington, DC 20552, (202) 377-6000, Chairman: Robert H. McKinney.

Generates and administers regulations governing the financing of solar equipment and solar-equipped homes.

Small Business Administration

1441 L Street NW, Washington, DC 20416, (202) 653-6365, Administrator: A. Vernon Weaver, Jr.

Provides financial and management assistance to businesses related to solar energy and conservation; administers special loan fund for solar enterprises.

Smithsonian Institution Radiation Biology Laboratory

12441 Parklawn Drive, Rockville, MD 20852, Director: William H. Klein.

Specializes in measurement of distribution of solar energy received at earth's surface and of solar constant values; instrumentation for measuring ultraviolet.

National Research Laboratories

A number of large research facilities in the United States are supported by research contracts from various departments and agencies of the U.S. Government; they are often administered and staffed by university personnel. The following National Laboratories are active in solar research:

Argonne National Laboratory, 9700 South Cass Avenue, Argonne, IL 60439

Battelle Pacific Northwest Laboratories, P.O. Box 999, Richland, WA 99352

Brookhaven National Laboratory, Upton, Long Island, NY 11973

Fermi National Accelerator Laboratory, P.O. Box 500, Batavia, IL 60601

Lawrence Berkeley Laboratory, 1 Cyclotron Road, Berkeley, CA 94720

Lawrence Livermore Laboratory, P.O. Box 808, Livermore, CA 94550

Los Alamos Scientific Laboratory, Mail Stop 571, Los Alamos, NM 87545

Oak Ridge National Laboratory, Box Y, Building 9102, Oak Ridge, TN 37830

Sandia Laboratories, Albuquerque, NM 87115

Congress

Congressional Committees

The United States Senate, Washington, DC 20510, or

The United States House of Representatives, Washington, DC 20515, (202) 224-3121 for both.

Congressional committees do the work of preparing and considering legislation. They maintain permanent staffs to do research, prepare position papers, help draft legislation, maintain liaison with other offices, and generally administer committee affairs. As part of their responsibilities, the following committees deal with energy conservation, renewable energy resources, and-or development of solar technologies.

Senate

Committee on Energy and Natural Resources

Chairman: Sen. Henry M. Jackson (D – WA)

Committee on Banking, Housing, and Urban Affairs
Chairman: Sen. William Proxmire (D – WI)

Committee on Governmental Affairs — Subcommittee on Energy, Nuclear Proliferation, and Federal Service
Chairman: Sen. John Glenn (D – OH)

House of Representatives

Committee on Interior and Insular Affairs — Subcommittee on Energy and the Environment
Chairman: Rep. Morris K. Udall (D – AZ)

Committee on Science and Technology — Subcommittee on Energy and Development
Chairman: Rep. Richard L. Ottinger (D – NY)

Committee on Government Operations — Subcommittee on Environment, Energy, and Natural Resources
Chairman: Rep. Anthony Toby Moffett (D – CT)

Committee on Interstate and Foreign Commerce — Subcommittee on Energy and Power
Chairman: Rep. John D. Dingell (D – MI)

Committee on Banking, Housing, and Urban Affairs
Chairman: Rep. Henry S. Reuss (D – WI)

Joint Committees and Offices

Joint Economic Committee — Subcommittee on Energy
Chairman: Sen. Edward M. Kennedy (D – MA)

Office of Technology Assessment
600 Pennsylvania Avenue SE, Washington, DC 20510.

The office maintains a staff to provide congressional committees with assessments or studies that identify the broad range of consequences, social as well as physical, that can be expected to evolve from various policy choices relating to the uses of technologies.

The Congressional Solar Coalition is composed of senators and representatives who initiate and support legislation that promotes solar energy and energy conservation. The members of the coalition, as of May 1979, were:

Senate

Sen. John L. H. Chaffee (R – RI)
Sen. Mark O. Hatfield (R – OR)
Sen. H. John Heinz, III (R – PA)
Sen. Patrick J. Leahy (D – VT)
Sen. Spark M. Matsunaga (D – HI)
Sen. Charles H. Percy (R – IL)
Sen. Abraham Ribicoff (D – CT)
Sen. Paul S. Sarbanes (D – MD)
Sen. Jim Sasser (D – TN)
Sen. Paul E. Tsongas (D – MA)

House of Representatives

Rep. Les AuCoin (D – OR)
Rep. Alvin Baldus (D – WI)
Rep. Max Baucus (D – MT)
Rep. Anthony Beilenson (D – CA)
Rep. Douglas R. Berenter (R – NE)
Rep. James J. Blanchard (D – MI)
Rep. Don Bonker (D – WA)
Rep. James C. Cleveland (R – NH)
Rep. Tony Coelho (D – CA)
Rep. Dan Daniel (R – VA)
Rep. Joel Deckard (R – IN)
Rep. Norman Dicks (D – WA)
Rep. Robert F. Drinan (D – MA)
Rep. John G. Fary (D – IL)
Rep. Dante Fascall (D – FL)
Rep. Vic Fazio (D – CA)
Rep. Martin Frost (D – TX)
Rep. Frank J. Gaurini (D – NJ)
Rep. Newt Gingrich (R – GA)
Rep. Dan Glickman (D – KS)
Rep. S. William Green (R – NY)
Rep. Kent Hance (D – TX)
Rep. James J. Howard (D – NJ)
Rep. Jerry Huckaby (D – LA)
Rep. James M. Jeffords (R – VT)
Rep. Dale E. Kildee (D – MI)
Rep. Jim Lloyd (D – CA)
Rep. Clarence D. Long (D – MD)
Rep. Ron Marlenee (R – MT)
Rep. Robert T. Matsui (D – CA)
Rep. Norman Y. Mineta (D – CA)
Rep. Steve Neal (D – NC)
Rep. Jerry M. Patterson (D – CA)
Rep. Donald J. Pease (D – OH)
Rep. Claude Pepper (D – FL)
Rep. Peter Peyser (D – NY)
Rep. J. J. Pickle (D – TX)
Rep. Patricia Schroeder (D – CO)
Rep. F. James Sensenbrenner, Jr. (R – WI)
Rep. Philip R. Sharp (D – IN)
Rep. Paul Simon (D – IL)
Rep. Gladys Spellman (D – MD)
Rep. Tom Steed (D – OK)
Rep. Gerry E. Studds (D – MA)
Rep. Bruce F. Vento (D – MN)
Rep. Harold L. Volkmer (D – MO)
Rep. James Weaver (D – OR)
Rep. G. William Whitehurst (R – VA)
Rep. Timothy E. Wirth (D – CO)
Rep. Leslie L. Wolff (D – NY)
Rep. Howard Wolpe (D – MI)
Rep. Robert Young (D – MO)

The New England Congressional Caucus, 53 D Street, SE, Washington, DC 20003, is a bipartisan organization composed of all 25 members of the House of Representatives from the six New England states. The caucus has an Energy Task Force and supports solar legislation.

Members of the Senate or House of Representatives may be reached at:

The United States Senate
Washington, DC 20510, or

The United States House of Representatives
Washington, DC 20515

Private Corporations

The following reproduction of the *Third Annual Deskbook Directory of Solar Product Manufacturers** shows manufacturers of all types of solar products, including those related to the use of solar energy in buildings. Its organization is in table format, including each manufacturer's total product line. Products appear under the following classifications:

absorbers and tubing
adhesives
air handlers
back-up systems
biomass systems and equipment
chemicals
  additves and heat transfer fluids
  storage
chillers
coatings and linings
  collector
  tank
collectors
  concentrating
  flat plate air
  flat plate liquid
  solar electric-solar thermal
  swimming pool
  vacuum tube
controls
  air systems controls
  differential thermostats
  packages
  packaged pumping and control modules
  packaged pumping, control and heat
    transfer modules
  tracking
  other
design aids
energy management systems
flow measurement devices
glazing
  collector
  passive
greenhouses
heat exchangers
heat pumps
heat recovery systems and components
housing and framing, collector
instrumentation
  radiation
  wind
  other weather
insulation
  collector
  passive
  tank-piping
irrigation systems
mounting hardware, collector
passive components
  glazing
  movable insulation and shading devices
  swimming pool covers
  other
passive systems
photovoltaics
pipes, fitting, piping specialties
power generation
  components
  systems
pumps
Rankine engines
reflectors
sealants, adhesives, gaskets
sensors
solar engines

The Solar Engineering Master Catalog (see Part IV, "Directories") carries more detailed information about solar companies.

storage bins
swimming pool covers
systems
  combined
  cooling
  distillation units
  heat pumps
  hot water
  greenhouses
  passive
  power generation
  space heating, air
  space heating, liquid
  swimming pool
  wind generation
tanks
  domestic water heater
  expansion
  storage
tools, equipment
valves
wind generation
  components
  systems

Collectors and collector components

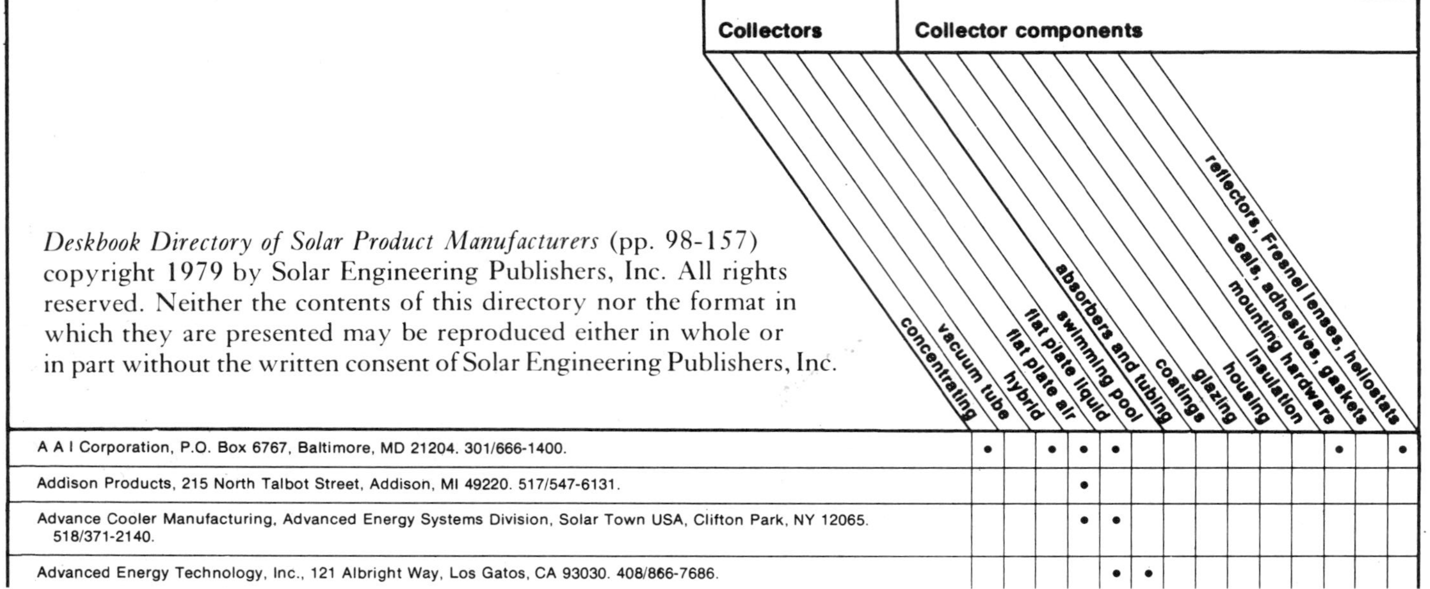

	Collectors						Collector components							
	concentrating	vacuum tube	hybrid	flat plate air	flat plate liquid	swimming pool	absorbers and tubing	coatings	glazing	housing	insulation	mounting hardware	seals, adhesives, gaskets	reflectors, Fresnel lenses, heliostats
A A I Corporation, P.O. Box 6767, Baltimore, MD 21204. 301/666-1400.	•		•	•	•							•		•
Addison Products, 215 North Talbot Street, Addison, MI 49220. 517/547-6131.				•										
Advance Cooler Manufacturing, Advanced Energy Systems Division, Solar Town USA, Clifton Park, NY 12065. 518/371-2140.				•	•									
Advanced Energy Technology, Inc., 121 Albright Way, Los Gatos, CA 93030. 408/866-7686.					•	•								

Aerolix Corporation, 720 South Columbus Avenue, Mt. Vernon, NY 10550. 914/699-1054.								•						
AFG Industries, P.O. Box 929, Kingsport, TN 37662. 615/245-0211.									•					
Air Comfort, Inc., 1001 Trinity Road, Raleigh, NC 27607. 919/851-2201.				•	•	•	•							
Aircraftsman, P.O. Box 628, Millbrook, AL 36054. 205/285-4469.					•									
Airmax, Inc., P.O. Box 129, White Oak, TX 75693. 214/836-2785.							•							
All Sun Power, Inc., 10400 South West 187th, Miami, FL 33157. 305/233-2224.					•	•	•							
Alpha Solarco, 1014 Vine Street, Suite 2230, Cincinnati, OH 45202. 513/621-1243.	•				•									
Alten Corporation, 2594 Leghorn Street, Mountain View, CA 94043. 415/969-6474.					•	•								
Alternate Energy Sources, 752 Duvall, Salina, KS 67401. 913/825-8218.					•									
Alternative Energy Resources, Inc., 1155 Larry Mahan Drive, El Paso, TX 79925. 915/593-1927.				•	•									
American Home Solar Energy Systems, 23142 Alcalde, Laguna Hills, CA 92653. 714/951-8507.					•	•								
American Klegecell Corporation, 204 North Dooley Street, Grapevine, TX 76051. 817/481-3547.										•	•		•	
American Solar King Corporation, 6801 New McGregor Highway, Waco, TX 76710. 817/776-3860.		•		•	•									
American Solar Products, Division of Squires Laboratories, 1996 South Valley Drive, Las Cruces, NM 88001. 505/523-5459.	•			•	•	•	•							
Ametek Power Systems Group, 1025 Polinski Road, Ivyland, PA 18974. 215/441-8770.					•									
Amicks Solar Heating, 375 Aspen Street, Middletown, PA 17057. 717/944-4544.					•									
Approtech, 770 Chestnut Street, San Jose, CA 95110. 408/297-6527.					•	•								
Aquasolar, Inc., 1251 Seeds Avenue, Sarasota, FL 33577. 813/366-7080.						•								
Aton Solar Manufacturing, 20 Pamaron, Novato, CA 94947. 415/883-0866.	•													
Automated Building Components, Inc., P.O. Box 592037, Miami, FL 33159. 800/327-3081.							•							
Automatic Power, Inc., 213 Hutcheson Street, Houston, TX 77003. 713/228-5208.	•													
Aztec Solar Company, P.O. Box 272, Maitland, FL 32751. 305/628-5004.					•									
Beam Engineering, Inc., 732 North Pastoria Avenue, Sunnyvale, CA 94086. 408/738-4573.	•													
Beeman Industries, South Road, East Hartland, CT 06024. 203/653-3073.							•							

Collectors/collector components

	Collectors						Collector components							
	concentrating	vacuum tube	hybrid	flat plate air	flat plate liquid	swimming pool	absorbers and tubing	coatings	glazing	housing	insulation	mounting hardware	seals, adhesives, gaskets	reflectors, Fresnel lenses, heliostats
Bell Industries Solar Division, 740 East 111th Place, Los Angeles, CA 90059. 213/777-2154.					•	•	•				•			
Berry Solar Products, Woodbridge at Main, P.O. Box 327, Edison, NJ 08817. 201/549-3800.	•						•	•		•		•		•
Bio-Energy Systems, Inc., Box 87, Ellenville, NY 12428. 914/647-6482.				•	•	•	•			•			•	
Brown Manufacturing Company, P.O. Box 14546, Oklahoma City, OK 73114. 405/751-1323.	•													
California Sun Energy, P.O. Box 732, Sunnymead, CA 92388. 714/653-4176.					•	•	•					•		
Calmac Manufacturing Corporation, 150 South Van Brunt Street, Englewood, NJ 07631. 201/569-0420.						•								
CE Glass, Combustion Engineering, Inc., 825 Hylton Road, Pennsauken, NJ 08110. 609/662-0400.									•					
Cities Service Company, New Haven Copper Operations, 79 Main Street, Seymour, CT 06483. 203/888-2551.							•							
Cole Solar Systems, Inc., 440-A East Saint Elmo Road, Austin, TX 78745. 512/444-2565.					•	•								
Colt Inc., 71590 San Jacinto, Rancho Mirage, CA 92270. 714/346-8033.					•	•								
Columbia Chase Solar Energy, 55 High Street, Holbrook, MA 02343. 617/767-0513.					•	•	•			•				

Conserdyne Corporation, 4437 San Fernando Road, Glendale, CA 91204. 213/246-8408.					•									
Contemporary Systems, CSI Solar Center, Route 12, Walpole, NH 03603. 603/756-4796.				•										
Copper State Solar Products, Inc., 4610 South 35th Street, P.O. Box 20504, Phoenix, AZ 85036. 602/276-4221.					•	•				•				
Custom Solar Heating Systems Company, P.O. Box 375, Albany, NY 12201. 518/438-7358.				•										
CY-RO Industries; 697 Route 46, Clifton, NJ 07015. 201/546-7900.									•					
The Dampney Company, 85 Paris Street, Everett, MA 02149. 617/389-2805.								•						
Datron, 20700 Plummer Street, Chatsworth, CA 91311. 213/882-9616.	•	•								•		•		
Deltair Solar Systems, Inc., Route 2, Box 53D, Chaska, MN 55318. 612/442-2131.				•										
Dow Corning Corporation, 2200 West Salzburg Road, CO 2314, Midland, MI 48640. 517/496-4000.								•					•	
Dumont Industries, P.O. Box 117, Monmouth, ME 04259. 207/933-4811.					•									
Dyrelite Corporation, 63 David Street, New Bedford, MA 02741. 617/993-9955.											•			
Eastern Sun-Tech Industries, Inc., 100 Merrick Road, P.O. Box 434, Rockville Center, NY 11570. 516/764-5678.					•									
Eastman Chemical Products, Inc., Plastics Division, Kingsport, TN 37662. 615/247-0411.									•					
E.C.S. Inc., Box 77, Concord Road, RD 3, Lebanon, NJ 08833. 201/236-2976.	•				•									
E. I. DuPont Company, Flouropolymer Division, Plastics, Products and Resins Department, Wilmington, DE 19898. 302/772-5800.									•					
E & K Service Company, 16824 74th N.E., Bothell, WA 98011. 206/488-2863.					•									
Energy Alternatives, Inc., 2102 South Broadway, Wichita, KS 67211. 316/262-6483.				•	•									
Energy Control Systems, 3324 Octavia Street, Raleigh, NC 27606. 919/851-2310.				•										
Energy Design Corporation, P.O. Box 34294, Memphis, TN 38134. 901/382-3000.	•	•			•									
Energy Distribution Inc., P.O. Box 353 Snug Harbor, Duxbury, MA 02332. 617/878-6793.					•									
Energy Saver Systems, Inc., 140 South Main Street, Wilkes-Barre, PA 18701. 717/829-5390.				•	•	•	•							
Energy Solutions, Inc., U.S. Highway 93, P.O. Drawer J, Stevensville, MT 59870. 406/777-5640.					•	•						•		
Energy Systems, Inc., 4570 Alvarado Canyon Road, San Diego, CA 92120. 714/280-6660.					•	•	•							
Entropy, Ltd., 5735 Arapahoe Avenue, Boulder, CO 80303. 303/443-5103.	•					•						•		•

Collectors/collector components

	Collectors: concentrating	Collectors: vacuum tube	Collectors: hybrid	Collectors: flat plate air	Collectors: flat plate liquid	Collectors: swimming pool	Collector components: absorbers and tubing	Collector components: coatings	Collector components: glazing	Collector components: housing	Collector components: insulation	Collector components: mounting hardware	Collector components: seals, adhesives, gaskets	Collector components: reflectors, Fresnel lenses, heliostats
ERGENICS, 681 Lawlins Road, Wyckoff, NJ 07481. 201/891-9103.					•		•	•						
FAFCO, Inc., 235 Constitution Drive, Menlo Park, CA 94025. 415/321-3650.						•								
Ferro Corporation, One Erie View Plaza, Cleveland, OH 44114. 216/641-8580.								•						
Flexonics Division, UOP, Inc., 300 East Devon Avenue, Bartlett, IL 60103. 312/837-1811.							•							
Ford Motor Company, Glass Division, 1 Parklane Boulevard, Suite 1000E, Dearborn, MI 48126. 313/568-2300.									•					
Fred Rice Productions, Inc., P.O. Box 643, 48780 Eisenhower Drive, La Quinta, CA 92253. 714/564-4823.	•													
Futuristic Solar Systems Corporation, 4900 Beech Place, Temple Hills, MD 20031. 301/899-3430.					•	•								
Gem Manufacturing Corporation, 390 East Tiffin Street, Bascom, OH 44809. 419/937-2225.				•						•		•		•
General Electric Company, P.O. Box 13601, Philadelphia, PA 19101. 215/962-2112.		•												
General Energy Devices, Inc., 1753 Ensley Avenue, Clearwater, FL 33516. 813/536-3585.					•	•								
General Solar Systems Division, General Extrusions, Inc., 4040 Lake Park Road, #2687, Youngstown, OH 44507. 216/783-0270.	•			•			•			•		•		•

Goettl Air Conditioning, Inc., 2005 East Indian School Road, Phoenix, AZ 85016. 602/273-1466.					•	•								
Grumman Energy Systems, Inc., 4175 Veterans Memorial Highway, Ronkonkoma, NY 11779. 516/737-3777.					•	•								
Halstead & Mitchell, P.O. Box 1110, Scottsboro, AL 35768. 205/259-1212.					•									
Halstead Industrial Products, Division of Halstead Industries, Inc., P.O. Box 309, Wynne, AR 72396. 501/238-3201.											•			
The Harshaw Chemical Company, 1945 East 97th Street, Cleveland, OH 44106. 216/721-8300.								•						
H. A. Schlatter, Inc., P.O. Box 5545, Rockford, IL 61125. 815/874-9471.							•							
Heldor Associates, 300 Kuller Avenue, Clifton, NJ 07015. 201/546-2121.												•		
Heliodyne, Inc., 770 South 16th Street, Richmond, CA 94804. 415/237-9614.					•									
Heliosystems, Inc., 3407 Ross Avenue, Dallas, TX 75206. 214/824-5971.				•	•	•				•				
Helix Solar Systems, P.O. Box 2038, City of Industry, CA 91746. 213/330-3312.					•	•	•					•		
Highland Plating Company, 1128 North Highland Avenue, Los Angeles, CA 90038. 213/469-2289.								•						
Horizon Enterprises, Inc., 1011 Northwest 6th Street, Homestead, FL 33030. 305/245-5145.					•	•						•		
Hughson Chemicals, 2000 W. Grandview Blvd., P.O. Box 1099, Erie, PA 16512. 814/868-3611.								•					•	
Hyperion, Inc., 4860 Riverbend Road, Boulder, CO 80303. 303/449-9544.	•			•	•	•	•							
ILI, Inc., 5965 Peachtree Corners East, Atlanta, GA 30071. 404/449-5900.					•									
Illini Insulation & Sun Energy Conservation Experts, 906 South Cherry Street, P.O. Box 172, Effingham, IL 62401. 217/347-7935.					•									
IMEX Trading, Ltd., 3315 Como Lane, San Jose, CA 95118. 408/264-2591.					•									
Impac Corporation, 312 Blondeau Street, Keokuk, IA 52632. 319/524-3304.			•	•										
In Solar Systems, 455 Pettis Avenue, Mountain View, CA 94041. 415/964-2801.						•	•							
Insta-Foam Midwest, Inc., 8000 47th Street, Lyons, IL 60534. 312/447-9262.											•			
Intertechnology/Solar Corporation, 276 Broadview Avenue, Warrenton, VA 22186. 703/347-9500.	•		•		•	•								
Isophenol Division, Robert Mitchell Solar Systems Design, Route 3, Box 239, Selkirk, NY 12158. 518/767-3100.											•			
Jacobs-Del Solar Systems, Inc., 251 South Lake Avenue, Pasadena, CA 91101. 213/449-2171.	•													

Collectors/collector components

	Collectors: concentrating	Collectors: vacuum tube	Collectors: hybrid	Collectors: flat plate air	Collectors: flat plate liquid	Collectors: swimming pool	Collector components: absorbers and tubing	Collector components: coatings	Collector components: glazing	Collector components: housing	Collector components: insulation	Collector components: mounting hardware	Collector components: seals, adhesives, gaskets	Collector components: reflectors, Fresnel lenses, heliostats
Jamak, Inc., 1401 North Bowie Drive, Weatherford, TX 76086. 817/594-8771.							•				•		•	
J. Catalano & Sons, Inc., 301 Stagg Street, Brooklyn, NY 11206. 212/821-6100.					•				•					
J. C. Solar, 3253 West Saginaw, Fresno, CA 93711. 209/227-4084.					•									
Jem-Sol, Inc., 409 North Grand Avenue, East, Springfield, IL 62702. 217/753-8330.					•									
Jet Air, Inc., Solar Heating Division, 1071 Industrial Place, El Cajon, CA 92020. 714/286-3550.					•	•								
John M. Frey Company, P.O. Box 560, 530 34th Street, Moline, IL 61265. 309/797-1161.					•						•			
Kalwall Corporation, P.O. Box 237, Manchester, NH 03105. 603/627-3861.				•			•	•	•					
Kee Klamps, 79 Benbro Drive, Buffalo, NY 14225. 716/685-1250.												•		
Kennard Industries, Inc., 8229 Brentwood Industrial Drive, St. Louis, MO 63144. 314/781-8013.					•									
Kingston Industries Corporation, 205 Lexington Avenue, New York, NY 10016. 212/889-0190.														•
Kirkhill Rubber Company, 300 East Cypress Street, Brea, CA 92621. 714/529-4901.											•		•	

KTA Products Division, NPD Energy Systems, Inc., 1455 Research Boulevard, Rockville, MD 20850. 301/424-4870.					•	•	•							
Largo Solar Systems, Inc., 991 South 40th Avenue, Ft. Lauderdale, FL 33317. 305/583-8090.					•		•					•		
Lasco Industries, 3255 East Miraloma, Anaheim, CA 92806. 714/993-1220.									•	•				
Lennox Industries, Inc., P.O. Box 400450, Dallas, TX 75240. 214/783-5427.					•		•							
Libbey Owens Ford Company, 1701 East Broadway, Toledo, OH 43605. 419/247-4350.					•	•								
The Lord's Power Company, L.P.C., Inc., P.O. Box 458, Nora Springs, IA 50458. 515/749-2528.				•			•			•				
Mann-Russell Electronics, Inc., 1401 Thorne Road, Tacoma, WA 98421. 206/383-1591.														•
Metco, Inc., 101 Prospect Avenue, Westbury, NY 11590. 516/334-1300.								•						
Microweld, Inc., 165 Brook Street, Franklin, MA 02038. 617/528-5611.							•							
Mid-Western Solar & Insulation Inc., 2235 Irvin Cobb Drive, Paducah, KY 42001. 502/443-6295.				•			•							
MNK Enterprises, Inc., P.O. Box 87, Bancroft, ID 83217. 208/648-7668.					•									
Mobay Chemical Corporation, Plastics and Coatings Division, Parkway West, Pittsburgh, PA 15205. 412/777-2000.									•					
Mor-Flo Industries, Inc., 18450 South Miles, Cleveland, OH 44128. 216/663-7300.						•								
Mueller Brass Company, 1925 Lapeer Avenue, Port Huron, MI 48060. 313/987-4000.							•							
Multi-Research Corporation, Highway 35, Keyport, NJ 07735. 201/264-8000.					•									
National Metallizing, A Saxon Industries Company, P.O. Box 5202, Princeton, NJ 08540. 800/257-5116.									•		•			
National Solar Supply Company, 2331 Adams Drive Northwest, Atlanta, GA 30318. 404/352-3478.					•									
Natural Energy Resources, 840 East Bethany Home, Phoenix, AZ 85014. 602/266-2949.					•	•	•							
Naturgy, P.O. Box 3403, Tulsa, OK . 918/587-7175.					•									
New Jersey Aluminum, P.O. Box 73, New Brunswick, NJ 08902. 800/631-5856.							•			•				
Nor-Ell, Inc., 851 Hubbard Avenue, St. Paul, MN 55104. 612/487-1441.								•						
Norton Company, Sealants Division, Interstate 290, Industrial Center, Northboro, MA 01532. 617/393-6064.													•	
Novan Energy, Inc., 3163 Walnut Street, Boulder, CO 80301. 303/447-9193.					•									
NPD Solarcrome, 333 South Teilman, Fresno, CA 93706. 209/442-1773.								•						

Collectors/collector components

	Collectors						Collector components							
	concentrating	vacuum tube	hybrid	flat plate air	flat plate liquid	swimming pool	absorbers and tubing	coatings	glazing	housing	insulation	mounting hardware	seals, adhesives, gaskets	reflectors, Fresnel lenses, heliostats
OEM Products, Inc., 2701 Adamo Drive, Tampa, FL 33605. 813/247-5947.					•	•								
Olin Brass, East Alton, IL 62024. 618/258-2443.					•	•	•							
Olympic Solar Corporation, 208 15th Street, Southwest, Canton, OH 44707. 216/452-8856.								•						
Omnium-G, 1815.5 North Orange-Thorpe Park, Anaheim, CA 92804. 714/879-8421.	•		•											•
Otto Fabric, Inc., P.O. Box 18361, Wichita, KS 67218. 316/262-5183.							•		•					
Overly Manufacturing Company, 2943 Gleneden, Los Angeles, CA 90039. 213/663-8161.					•	•								
Owens-Illinois, Inc., P.O. Box 1035, Toledo, OH 43601. 419/247-9099.		•												
Park Energy Company, Star Route 9, Jackson, WY 83001. 307/733-4950.				•			•		•					
Pawling Rubber Corporation, 157 Maple Boulevard, Pawling, NY 12564. 914/855-1000.													•	
Permaloy Corporation, P.O. Box 1559, Ogden, UT 84402. 801/731-4303.				•	•		•							
Phelps Dodge Brass Company, Solar Division, 1050 Wall Street West, Lyndhurst, NJ 07071. 201/935-8180.						•	•							

Photometric Design Enterprises, 1107 Berkshire Drive, Westland, MI 48185. 313/729-7044.	•													
Pittsburgh Corning Corporation, 800 Presque Isle Drive, Pittsburgh, PA 15239. 412/327-6100.											•			
Porter Energy Products, P.O. Box 827, Newark, DE 19711. 301/398-0284.					•									
Practical Solar Systems, 7221 South 180th Street, Kent, WA 98031. 206/226-4554.				•	•									
Radco Products, Inc., 2877 Industrial Parkway, Santa Maria, CA 93454. 805/928-1881.					•	•	•			•				
Ra-Energy Systems, Inc., 11459 Woodside Avenue, Lakeside, CA 92040. 714/448-6216.					•	•	•							
Raypak, Inc., 31111 Agoura Road, Westlake Village, CA 91360. 213/889-1500.					•	•								
Refrigeration Research, Solar Research Division, 525 North Fifth Street, Brighton, MI 48116. 313/277-1151.					•		•	•				•		
Research Products Corporation, P.O. Box 1467, 1015 East Washington Avenue, Madison, WI 53713. 608/257-8801.				•										
Revere Solar and Architectural Products, Inc., P.O. Box 151, Rome, NY 13440. 315/338-2595.					•	•								
Rheem/Ruud Water Heater Divisions, City Investing Company, 5780 Peachtree-Dunwoody Road, Northeast, Suite 400, Atlanta, GA 30342. 404/252-7211.					•							•		
Rigidized Metals Corp., 658 Ohio Street, Buffalo, NY 14203. 716/849-4702.							•							•
R. J. Sullivan & Sun, 1835 3rd Avenue, Southeast, Cedar Rapids, IA 52403. 319/366-0030.	•													
R-M Products, 5010 Cook, Denver, CO 80216. 303/825-0203.				•	•									
Roll Forming Corporation, Industrial Park, Shelbyville, KY 40065. 502/633-4435.										•				
Rom-Aire Solar Systems, 121 Miller Road, Avon Lake, OH 44012. 216/933-5000.				•										
Ryniker Steel Products Company, P.O. Box 1932, Billings, MT 59103. 406/252-3836.				•										
Sebra Solar Energy, 500 North Tucson Boulevard, Tucson, AZ 85716. 602/881-6555.				•		•								
SETSCO, 1037-B Shary Circle, Concord, CA 94518. 415/676-5392.					•	•								
Sheffield Plastics, Inc., Salisbury Road, Sheffield, MA 01257. 413/229-8711.									•					
Shelley Radiant Ceiling Company, 456 West Frontage Road, Northfield, IL 60093. 312/446-2800.							•	•						
Silicon Sensors, Inc., Highway 18, East, Dodgeville, WI 53533. 608/935-2707.			•											
SJC Corporation, Division Frigiking-Tappan, 206 Woodford Avenue, Elyria, OH 44036. 216/329-2000.					•	•							•	
SMC Energy Company, P.O. Box 246, Omaha, NE 68102. 402/397-1370.				•	•									

Collectors/collector components

	Collectors						Collector components							
	concentrating	vacuum tube	hybrid	flat plate air	flat plate liquid	swimming pool	absorbers and tubing	coatings	glazing	housing	insulation	mounting hardware	seals, adhesives, gaskets	reflectors, Fresnel lenses, heliostats
Solafern, Ltd., 536 MacArthur Boulevard, Bourne, MA 02532. 617/563-7181.				•										
Solara Associates, Inc., 1001 Connecticut Avenue, Northwest #632, Washington, DC 20036. 202/296-7070.	•				•	•								
Solar Alternative, Inc., 22 South Main Street, Brattleboro, VT 05301. 802/254-6668.					•	•	•		•					
Solaray Corporation, 2414 Makiki Heights Drive, Honolulu, HI 96822. 808/533-6464.							•							
Solar by WEFCO, Inc., 324 South Kidd Street, Whitewater, WI 53190. 414/473-2525.				•	•									
Solar Contact Systems, 1415 Vernon, Anaheim, CA 92805. 714/991-8120.					•							•		
Solar Development, Inc., 3630 Reese Avenue, Riviera Beach, FL 33404. 305/842-8935.				•	•	•	•					•		
Solar Development, Inc. Northwest, 690 Yellowstone, Suite D, Pocatello, ID 83201. 208/233-6563.				•	•	•						•		
Solar Dynamics Corporation, 550 Frontage Road, Northfield, IL 60093. 312/446-5242.					•									
Solar Dynamics of Arizona, 1100 North Lake Havasu Avenue, Suite H, Lake Havasu City, AZ 86403. 602/855-5051.					•	•	•					•		
Solar-En Corporation, 3118 Route 10, Denbrook Village, Denville, NJ 07834. 201/361-2300.				•	•	•								

Company														
Solar Energies of California, 11421 Woodside Avenue, Lakeside, CA 92040. 714/448-4300.				•	•	•								
Solar Energy Components, Inc., 212 Welsh Pool Road, Lionville, PA 19353. 215/644-9017.						•								
Solar Energy Engineering, 31 Maxwell Court, Santa Rosa, CA 95401. 707/542-4498.					•	•	•							
Solar Energy of the South, 3266 International Drive, Mobile, AL 36606. 205/478-3647.					•	•								
Solar Energy Products, Inc., Mountain Pass, Hopewell Junction, NY 12533. 914/226-8596.					•	•	•	•		•		•		
Solar Energy Research Corporation, 1224 Sherman Drive, Longmont, CO 80501. 303/772-8406.				•	•	•						•		
Solar Engines, 2937 West Indian School Road, Phoenix, AZ 85017. 602/274-3541.	•													•
Solar Enterprises, P.O. Box 1046, Red Bluff, CA 96080. 916/527-0551.					•	•								
Solar Enterprises, Inc., 7928 N.E. Main Street, Fridley, MN 55432. 612/784-7177.					•									
Solar Equipment Corporation, P.O. Box 357, Lakeside, CA 92040. 714/561-4531.	•		•				•							•
Solar Equipment Distributors, Inc., Division of Yago Systems Design, P.O. Box 64, Barboursville, WV 25504. 304/736-4091.												•		
Solar-Eye Products, Inc., 1300 N.W. McNab, Fort Lauderdale, FL 33309. 305/974-2500.				•	•		•			•				
Solar Farm Industries, Inc., P.O. Box 242, Stockton, KS 67669. 913/425-6726.				•			•							
Solar Fin Systems, 140 South Dixie Highway, St. Augustine, FL 32084. 904/824-3522.					•		•							
Solargenics, Inc., 20319 Nordhoff Street, Chatsworth, CA 91311. 213/998-0806.			•		•	•								
Solargizer International, P.O. Box 20142, Minneapolis, MN 55420. 612/944-7861.				•	•	•								
Solar Heat Company, P.O. Box 110, Greenville, PA,				•		•	•					•		
Solar Heat Corporation, 398 East 271st Street, Euclid, OH 44132. 216/732-8034.				•										
Solar Home Systems, Inc., 8732 Camelot, Chesterland, OH 44026. 216/729-9350.				•	•		•							
Solar Industries, Monmouth Airport Industrial Park, Farmingdale, NJ 07727. 201/938-7000.					•	•								
Solar Industries, Inc., Box 303, Plymouth, CT 06782. 203/283-0223.	•	•			•									•
Solar Innovations, 412 Longfellow Boulevard, Lakeworth, FL 33801. 813/688-8373.					•							•		
Solar Kinetics, Inc., P.O. Box 47045, Dallas, TX 75247. 214/630-9328.	•													
Solar King International, 8577 Canoga Park, Canoga Park, CA 91304. 213/998-6400.					•									

Collectors/collector components

	Collector components: reflectors, Fresnel lenses, heliostats	seals, adhesives, gaskets	mounting hardware	insulation	housing	glazing	coatings	absorbers and tubing	Collectors: swimming pool	flat plate liquid	flat plate air	hybrid	vacuum tube	concentrating
Solarlife, 404 Lippincott Avenue, Riverton, NJ 08077. 609/829-7022.													•	
Solar Living, Inc., P.O. Box 12, Netcong, NJ 07857. 201/691-8483.					•			•	•	•				
Solarnetics Corporation, 1654 Pioneer Way, El Cajon, CA 92020. 714/519-7122.									•	•				
Solarom of North America, Inc., 10 Central Avenue, Westbury, NY 11590. 516/334-1660.										•				
Solaron Corporation, 720 South Colorado Boulevard, Denver, CO 80222. 303/759-0101.			•					•	•	•	•			
Solar One, Ltd., 2644 Barrett Street, Virginia Beach, VA 23452. 804/340-7774.									•	•				
Solar Products, Inc., 614 Northwest 62nd Street, Miami, FL 33150. 305/756-7609.										•				
Solar Research Systems, 3001 Red Hill Avenue, I-105, Costa Mesa, CA 92626. 714/545-4941.			•						•					
Solar Servants, inc., 289 Tropical Shore Way, Fort Myers Beach, FL 33931. 813/463-5016.								•		•				
Solarsmith, 7613 Hendrix, Northeast, Albuquerque, NM 87110. 505/294-6522.										•				
Solar Southwest, 5700 Andrews Highway, Odessa, TX 79762. 915/362-4922.											•			

Solar Specialties, Inc., Route 7, P.O. Box 409, Golden, CO 80401. 303/642-3063.					•									
Solar Systems, Inc., 507 West Elm Street, Tyler, TX 75702. 214/592-5343.		•			•									
Solar Technology International, 119 North Center Street, Statesville, NC 28677. 704/873-7959.				•	•		•			•				
Solartec, Inc., 250 Pennsylvania Avenue, Salem, OH 44460. 216/332-9100.					•									
Solartherm, Inc., 1110 Fidler Lane, Silver Spring, MD 20910. 202/882-4000.				•	•	•								
Solar Thermal Systems, Division of Exxon Enterprises, Inc., 90 Cambridge Street, Burlington, MA 01803. 617/272-8460.					•	•	•							
Solar II Enterprises, 41 Lost Lake Lane, Campbell, CA 95008. 408/866-6244.					•	•	•							
Solar Unlimited, Inc., 204 Oakwood Avenue, Huntsville, AL 35811. 205/534-0661.					•	•	•					•		
Solar Warehouse, Inc., 140 Shrewsbury Avenue, Red Bank, NJ 07701. 201/842-2210.												•		
Solar Water Heaters of New Port Richey, Inc., 1214 U.S. 19 North, New Port Richey, FL 33552. 813/848-2343.					•									
Solcoor, Inc., 849 South Broadway 208, Los Angeles, CA 90014. 213/622-4181.					•	•	•							
Solecon, P.O. Box 09763, 2770 East Main Street, Columbus, OH 43209. 614/236-5982.				•			•							
Solectro-Thermo, 1934 Lakeview Avenue, Dracut, MA 01826. 617/957-0028.			•											
Solex, 187 Billerica Road, Chelmsford, MA 01824. 617/256-8724.						•								
Solpower Industries, Inc., 10211-C Bubb Road, Cupertino, CA 95014. 408/996-3222.					•	•	•							
Southeastern Solar Systems, Inc., 4705J Bakers Ferry Road, Atlanta, GA 30336. 404/691-1960.					•	•						•		
Spectran/Instruments, Inc., P.O. Box 891, La Habra, CA 90631. 213/694-3995.														•
S S Solar Inc., 16 Keystone Avenue, River Forest, IL 60305. 312/771-1912.	•													
S-Systems, Inc., Dallas, TX 75220. 214/358-4101.					•	•								
Standard Solar Collectors, Inc., 1465 Gates Avenue, Brooklyn, NY 11227. 212/456-1882.	•			•	•									
Sun-Bank, Inc., 924 North Main, Wichita, KS 67203. 316/265-0866.					•	•								
Sunburst Solar Energy, Inc., P.O. Box 2799, Menlo Park, CA 94025. 415/327-8022.					•	•	•							
Sun Craft, 5001 East 59th Street, Kansas City, MO 64130. 816/333-2100.				•										
The Sundu Company, 3319 Keys Lane, Anaheim, CA 92804. 714/828-2873.					•	•								

Collectors/collector components

	Collectors						Collector components							
	concentrating	vacuum tube	hybrid	flat plate air	flat plate liquid	swimming pool	absorbers and tubing	coatings	glazing	housing	insulation	mounting hardware	seals, adhesives, gaskets	reflectors, Fresnel lenses, heliostats
Sunduit, Inc., 281 East Jackson Street, Virden, IL 62690. 217/965-3352.					•		•			•				
Sunearth Solar Products Corporation, Route 1, Box 337, Green Lane, PA 18054. 215/699-7892.					•		•							
Sunflower Energy Works, Inc., P.O. Box 85, Goessel, KS 67053. 316/367-2647.				•			•							
Sun-Heet, Inc., 2624 South Zuni, Englewood, CO 80110. 303/777-2964.	•													
Sunmaster Corporation, 12 Spruce Street, Corning, NY 14830. 607/937-5441.	•	•												
Sun-Pac, Inc., P.O. Box 8169, Alexandria, LA 71306. 318/445-6751.					•	•								
Sunplate, Inc., 328 North Hightower, Thomaston, GA 30286. 404/647-2531.							•							
Sunpower Systems Corporation, 510 South 52nd Street, Suite 101, Tempe, AZ 85281. 602/894-2331.	•													
Sun Power Systems, Ltd., 1024 West Maude Avenue, Suite 203, Sunnyvale, CA 94086. 408/738-2442.					•	•								
Sun-Ray Solar Equipment Company, Inc., 415 Howe Avenue, Shelton, CT 06484. 203/735-7767.	•			•	•	•	•							
Sun-Ray Solar Heaters, 4898 Ronson Court, San Diego, CA 92111. 714/278-7720.					•	•	•			•		•		

Sunray Solar Heat, Inc., 202 Classon Avenue, Brooklyn, NY 11205. 212/638-6540.					•	•								
Sunsav, Inc., 640 South Union Street, Lawrence, MA 01843. 617/687-0044.		•			•	•	•							
Sunshine Unlimited, 900 North Jay Street, Chandler, AZ 85224. 602/963-3878.					•	•								
Sunspot Environmental Energy Systems, P.O. Box 5110, San Diego, CA 92105. 714/264-9100.					•	•								
Sun Stone Solar Energy Equipment, P.O. Box 138, Baraboo, WI 53913. 608/356-7744.				•										
Sun Systems, Inc., P.O. Box 347, Boston, MA 02186. 617/265-9600.				•	•	•								
Suntree Solar Company, P.O. Box 1261, Woonsocket, RI 02895. 401/769-1689.	•				•	•	•		•	•		•		
Sunverter Company, Inc., Route 1, Box 269, Murphysboro, IL 62966. 618/687-3416.				•			•							
Sun Wise Inc., 609 10th Avenue South, P.O. Box 6622, Great Falls, MT 59406. 406/727-5977.				•	•							•		
Sunworks Division, Sun Solector Corp., P.O. Box 3900, Somerville, NJ 08876. 201/469-0399.				•	•		•							
Swedlow, Inc., 12122 Western Avenue, Garden Grove, CA 92645. 714/893-7531.														•
S. W. Energy Options, Route 8, Box 30-H, Silver City, NM 88061. 505/538-9598.				•					•					
Systems Technology, Inc., P.O. Box 337, Shalimar, FL 32579. 904/863-9213.					•									
T/Drill, Inc., 727 West Ellsworth, Building 8, Ann Arbor, MI 48104. 313/995-2187.							•							
Technitrek Corporation, 1999 Pike Avenue, San Leandro, CA 94577. 415/352-0535.					•	•						•		
Teledyne AERO-CAL, 528 East Mission Road, San Marcos, CA 92069. 714/744-1131.										•				•
Teledyne Metal Forming, 1937 Sterling Avenue, Box 757, Elkhart, IN 46515. 219/295-5525.							•			•				
Terra-Light, Inc., P.O. Box 493, Billerica, MA . .						•	•							
Texas Urethanes Inc., P.O. Box 9563, Austin, TX 78766. 512/272-5531.											•			
Thermatool Corp., 280 Fairfield Avenue, Stamford, CT 06902. 203/357-1555.							•							
3-E Corporation, 401 Kennedy Boulevard, Somerdale, NJ 08083. 609/784-8200.										•			•	
Universal Solar Development, Inc., 1633 Acme Street, Orlando, FL 32805. 305/423-8727.					•	•	•					•		
U.S. Solar Corporation, P.O. Drawer K, Hampton, FL 32044. 904/468-1517.					•	•						•		
Vegetable Factory Inc., 100 Court Street, Copaigue, NY 11726. 516/842-9300.				•		•								

Collectors/collector components

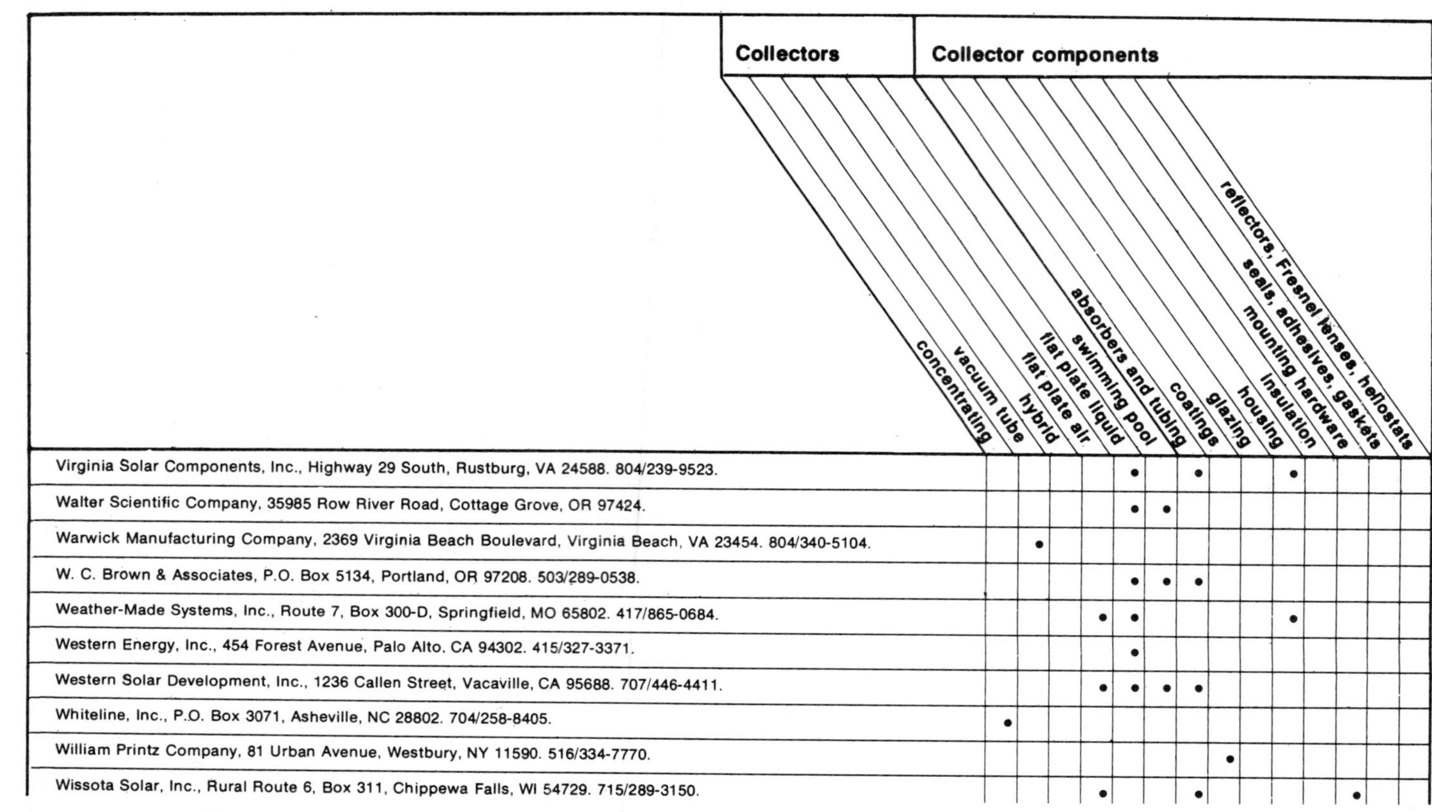

	Collectors						Collector components							
	concentrating	vacuum tube	hybrid	flat plate air	flat plate liquid	swimming pool	absorbers and tubing	coatings	glazing	housing	insulation	mounting hardware	seals, adhesives, gaskets	reflectors, Fresnel lenses, heliostats
Virginia Solar Components, Inc., Highway 29 South, Rustburg, VA 24588. 804/239-9523.					•		•			•				
Walter Scientific Company, 35985 Row River Road, Cottage Grove, OR 97424.					•	•								
Warwick Manufacturing Company, 2369 Virginia Beach Boulevard, Virginia Beach, VA 23454. 804/340-5104.		•												
W. C. Brown & Associates, P.O. Box 5134, Portland, OR 97208. 503/289-0538.					•	•	•							
Weather-Made Systems, Inc., Route 7, Box 300-D, Springfield, MO 65802. 417/865-0684.				•	•					•				
Western Energy, Inc., 454 Forest Avenue, Palo Alto, CA 94302. 415/327-3371.					•									
Western Solar Development, Inc., 1236 Callen Street, Vacaville, CA 95688. 707/446-4411.				•	•	•	•							
Whiteline, Inc., P.O. Box 3071, Asheville, NC 28802. 704/258-8405.	•													
William Printz Company, 81 Urban Avenue, Westbury, NY 11590. 516/334-7770.								•						
Wissota Solar, Inc., Rural Route 6, Box 311, Chippewa Falls, WI 54729. 715/289-3150.				•			•					•		

Ying Manufacturing Corporation, 1957 West 144st Street, Gardena, CA 90249 213/770-1756.	•			•	•									
Z. Z. Industries, 18 Spring Hill Terrace, Spring Valley, NY 10977. 914/356-9448.								•						
Z. Z. Corporation, 10806 Kaylor Street, Los Alamitos, CA 90720. 213/598-3220.	•	•												

Solar systems
Passive systems and components

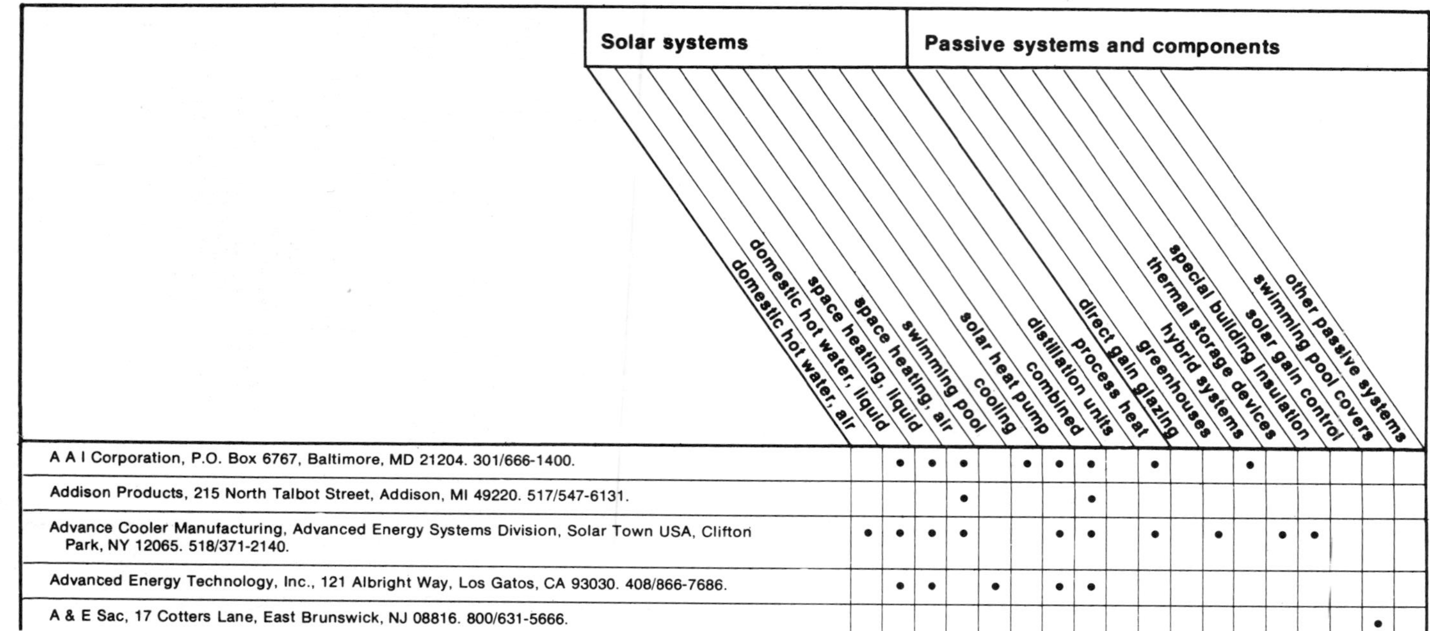

	Solar systems										Passive systems and components							
	domestic hot water, air	domestic hot water, liquid	space heating, liquid	space heating, air	swimming pool	cooling	solar heat pump	combined	distillation units	process heat	direct gain glazing	greenhouses	hybrid systems	thermal storage devices	special building insulation	solar gain control	swimming pool covers	other passive systems
A A I Corporation, P.O. Box 6767, Baltimore, MD 21204. 301/666-1400.		•	•	•		•	•	•		•			•					
Addison Products, 215 North Talbot Street, Addison, MI 49220. 517/547-6131.				•				•										
Advance Cooler Manufacturing, Advanced Energy Systems Division, Solar Town USA, Clifton Park, NY 12065. 518/371-2140.	•	•	•	•			•	•		•		•		•	•			
Advanced Energy Technology, Inc., 121 Albright Way, Los Gatos, CA 93030. 408/866-7686.		•	•		•		•	•										
A & E Sac, 17 Cotters Lane, East Brunswick, NJ 08816. 800/631-5666.																	•	

AFG Industries, P.O. Box 929, Kingsport, TN 37662. 615/245-0211.											•							
Air Comfort, Inc., 1001 Trinity Road, Raleigh, NC 27607. 919/851-2201.		•	•					•										
Aircon, Ltd., 2400 Florida Avenue, Norfolk, VA 23513. 804/853-7423.				•			•	•										
All Sun Power, Inc., 10400 South West 187th, Miami, FL 33157. 305/233-2224.		•			•													
Alpha Solarco, 1014 Vine Street, Suite 2230, Cincinnati, OH 45202. 513/621-1243.		•	•			•	•	•		•								
Alten Corporation, 2594 Leghorn Street, Mountain View, CA 94043. 415/969-6474.		•																
Alternate Energy Industries, 420 Lexington Avenue, Suite 1628, New York, NY 10017. 212/682-8220.		•																
Alternate Energy Sources, 752 Duvall, Salina, KS 67401. 913/825-8218.		•																
Alternative Energy Resources, Inc., 1155 Larry Mahan Drive, El Paso, TX 79925. 915/593-1927.	•	•																
American Klegecell Corporation, 204 North Dooley Street, Grapevine, TX 76051. 817/481-3547.															•			
American Solar King Corporation, 6801 New McGregor Highway, Waco, TX 76710. 817/776-3860.		•	•		•	•	•	•		•								
American Solar Products, Division of Squires Laboratories, 1996 South Valley Drive, Las Cruces, NM 88001. 505/523-5459.		•			•	•		•		•								
Ametek Power Systems Group, 1025 Polinski Road, Ivyland, PA 18974. 215/441-8770.		•																
Amicks Solar Heating, 375 Aspen Street, Middletown, PA 17057. 717/944-4544.		•						•										
A. O. Smith Corporation, P.O. Box 28, Kankakee, IL 60901. 815/933-8241.		•																
Approtech, 770 Chestnut Street, San Jose, CA 95110. 408/297-6527.	•	•	•		•		•	•	•					•				
Architectural Research Corporation, 13030 Wayne Road, Livonia, MI 48105. 313/525-9400.														•	•			
Arkla Industries, P.O. Box 534, Evansville, IN 47704. 812/424-3331.						•		•										
Aton Solar Manufacturing, 20 Pamaron, Novato, CA 94947. 415/883-0866.								•		•								
Beam Engineering, Inc., 732 North Pastoria Avenue, Sunnyvale, CA 94086. 408/738-4573.		•	•					•		•								
Bio-Energy Systems, Inc., Box 87, Ellenville, NY 12428. 914/647-6482.		•	•		•		•	•		•	•	•						
Brown Manufacturing Company, P.O. Box 14546, Oklahoma City, OK 73114. 405/751-1323.															•	•		
Buckmaster Industries, 23846 Sunnymead Boulevard, P.O. Box 730, Sunnymead, CA 92388. 714/653-8461.		•	•		•			•										

Solar systems/passive systems and components

	Solar systems: domestic hot water, air	domestic hot water, liquid	space heating, liquid	space heating, air	swimming pool	cooling	solar heat pump	combined	distillation units	Passive systems and components: process heat	direct gain glazing	greenhouses	hybrid systems	thermal storage devices	special building insulation	solar gain control	swimming pool covers	other passive systems
California Solar Systems Company, 532 Weddell Drive, Suite 4, Sunnyvale, CA 94086. 408/744-0651.	•	•																
California Sun Energy, P.O. Box 732, Sunnymead, CA 92388. 714/653-4176.		•	•		•		•	•		•		•						
Century Fiberglass, P.O. Box 6069, Anaheim, CA 92806. 714/630-0012.														•				
Cole Solar Systems, Inc., 440-A East Saint Elmo Road, Austin, TX 78745. 512/444-2565.		•	•		•			•										
Columbia Chase Solar Energy, 55 High Street, Holbrook, MA 02343. 617/767-0513.		•	•		•			•										
Conserdyne Corporation, 4437 San Fernando Road, Glendale, CA 91204. 213/246-8408.		•	•		•			•										
Contemporary Systems, CSI Solar Center, Route 12, Walpole, NH 03603. 603/756-4796.				•				•		•				•	•			
Copper State Solar Products, Inc., 4610 South 35th Street, P.O. Box 20504, Phoenix, AZ 85036. 602/276-4221.		•	•		•	•												
Cover Pools, Inc., 117 West Fireclay Avenue, Salt Lake City, UT 84107. 801/262-2724.																	•	
Custom Solar Heating Systems Company, P.O. Box 375, Albany, NY 12201. 518/438-7358.				•	•			•										

Company																		
CY-RO Industries, 697 Route 46, Clifton, NJ 07015. 201/546-7900.															●	●		
The Dampney Company, 85 Paris Street, Everett, MA 02149. 617/389-2805.																●		
Datron, 20700 Plummer Street, Chatsworth, CA 91311. 213/882-9616.								●										
Dearing Solar Energy Systems, 12324 Ventura Boulevard, Studio City, CA 91604. 213/769-2521.																	●	
Deltair Solar Systems, Inc., Route 2, Box 53D, Chaska, MN 55318. 612/442-2131.				●														
Dumont Industries, P.O. Box 117, Monmouth, ME 04259. 207/933-4811.		●																
Dunham-Bush, Inc., 101 Burgess Road, Harrisonburg, VA 22801. 703/434-0711.							●											
Dyrelite Corporation, 63 David Street, New Bedford, MA 02741. 617/993-9955.															●			
E. I. DuPont Company, Flouropolymer Division, Plastics, Products and Resins Department, Wilmington, DE 19898. 302/772-5800.											●					●		
E & K Service Company, 16824 74th N.E., Bothell, WA 98011. 206/488-2863.		●						●										
Energy Alternatives, Inc., 2102 South Broadway, Wichita, KS 67211. 316/262-6483.				●				●										
Energy Control Systems, 3324 Octavia Street, Raleigh, NC 27606. 919/851-2310.								●										
Energy Design Corporation, P.O. Box 34294, Memphis, TN 38134. 901/382-3000.								●		●								
Energy Distribution Inc., P.O. Box 353 Snug Harbor, Duxbury, MA 02332. 617/878-6793.		●										●						
The Energy Factory, 1550 North Clark, Fresno, CA 93703. 209/441-1833.				●				●										
Energy Management Engineering, Inc., 4241 Hogue Road, Evansville, IN 47712. 812/464-2463.			●					●										
Energy Solutions, Inc., U.S. Highway 93, P.O. Drawer J, Stevensville, MT 59870. 406/777-5640.		●	●		●			●										
Energy Systems, Inc., 4570 Alvarado Canyon Road, San Diego, CA 92120. 714/280-6660.		●																
Entropy, Ltd., 5735 Arapahoe Avenue, Boulder, CO 80303. 303/443-5103.		●	●		●	●	●	●	●	●								
ERGENICS, 681 Lawlins Road, Wyckoff, NJ 07481. 201/891-9103.					●											●		
EXSUN, P.O. Box 3745, Washington, DC 20007.															●	●		
FAFCO, Inc., 235 Constitution Drive, Menlo Park, CA 94025. 415/321-3650.					●													
FANJET, 6153 Crabtree Road, Columbia, SC 29206. 803/787-0243.																		●
Fiber-Rite Products, Inc., P.O. Box 38095, Cleveland, OH 44138. 216/235-6800.														●				

Solar systems/passive systems and components

Company	Solar systems: domestic hot water, air	domestic hot water, liquid	space heating, liquid	space heating, air	swimming pool	cooling	solar heat pump	combined	distillation units	process heat	Passive systems and components: direct gain glazing	greenhouses	hybrid systems	thermal storage devices	special building insulation	solar gain control	swimming pool covers	other passive systems
Ford Motor Company, Glass Division, 1 Parklane Boulevard, Suite 1000E, Dearborn, MI 48126. 313/568-2300.											•							
Four Seasons Solar Products Corporation, 672 Sunrise Highway, West Babylon, NY 11704. 516/422-1300.												•						
Fred Rice Productions, Inc., P.O. Box 643, 48780 Eisenhower Drive, La Quinta, CA 92253. 714/564-4823.															•	•		
Futuristic Solar Systems Corporation, 4900 Beech Place, Temple Hills, MD 20031. 301/899-3430.		•	•		•		•	•										
General Electric Company, P.O. Box 13601, Philadelphia, PA 19101. 215/962-2112.		•	•			•	•	•										
General Energy Devices, Inc., 1753 Ensley Avenue, Clearwater, FL 33516. 813/536-3585.		•	•	•	•			•		•								
General Solar Corporation, 51 Monroe Street, Rockville, MD 20850. 301/424-0545.																•		
General Solar Systems Division, General Extrusions, Inc., 4040 Lake Park Road, #2687, Youngstown, OH 44507. 216/783-0270.	•			•		•		•		•								
Goettl Air Conditioning, Inc., 2005 East Indian School Road, Phoenix, AZ 85016. 602/273-1466.		•			•		•											
The Graetz Group, 922 24th Street, North West, Suite 520, Washington, DC 20037. 202/466-4566.															•	•		

Green Mountain Homes, Waterman, Royalton, VT 05068. 802/763-8384.				•			•					•		•	•	•		
Grumman Energy Systems, Inc., 4175 Veterans Memorial Highway, Ronkonkoma, NY 11779. 516/737-3777.		•	•				•	•		•								
Halstead & Mitchell, P.O. Box 1110, Scottsboro, AL 35768. 205/259-1212.		•																
Heldor Associates, 300 Kuller Avenue, Clifton, NJ 07015. 201/546-2121.					•												•	
Heliodyne, Inc., 770 South 16th Street, Richmond, CA 94804. 415/237-9614.		•	•		•	•	•	•		•								
Heliosystems, Inc., 3407 Ross Avenue, Dallas, TX 75206. 214/824-5971.		•	•			•	•	•		•								
Helix Solar Systems, P.O. Box 2038, City of Industry, CA 91746. 213/330-3312.		•			•			•		•							•	
Holland Plastics, P.O. Box 248, Gilman, IA 50106. 515/498-7404.															•			
Horizon Enterprises, Inc., 1011 Northwest 6th Street, Homestead, FL 33030. 305/245-5145.		•						•										
Hyperion, Inc., 4860 Riverbend Road, Boulder, CO 80303. 303/449-9544.	•	•	•	•	•		•	•		•								
ILI, Inc., 5965 Peachtree Corners East, Atlanta, GA 30071. 404/449-5900.		•	•			•	•	•		•								
Illini·Insulation & Sun Energy Conservation Experts, 906 South Cherry Street, P.O. Box 172, Effingham, IL 62401. 217/347-7935.		•	•					•										
Intertechnology/Solar Corporation, 276 Broadview Avenue, Warrenton, VA 22186. 703/347-9500.		•	•		•		•	•		•								
Isophenol Division, Robert Mitchell Solar Systems Design, Route 3, Box 239, Selkirk, NY 12158. 518/767-3100.															•			
Jacobs-Del Solar Systems, Inc., 251 South Lake Avenue, Pasadena, CA 91101. 213/449-2171.			•			•	•	•		•								
J. Catalano & Sons, Inc., 301 Stagg Street, Brooklyn, NY 11206. 212/821-6100.															•	•		
Kalwall Corporation, P.O. Box 237, Manchester, NH 03105. 603/627-3861.	•			•				•			•			•	•			
Koolshade Corporation, P.O. Box 210, Solana Beach, CA 92075. 714/755-5126.																•		
Lamco, Inc., 5923 North Nevada Avenue, Colorado Springs, CO 80907. 303/599-8744.		•			•			•										
Largo Solar Systems, Inc., 991 South 40th Avenue, Ft. Lauderdale, FL 33317. 305/583-8090.		•							•									
Lasco Industries, 3255 East Miraloma, Anaheim, CA 92806. 714/993-1220.											•				•			
Lennox Industries, Inc., P.O. Box 400450, Dallas, TX 75240. 214/783-5427.		•	•			•		•		•								

Solar systems/passive systems and components

	Solar systems										Passive systems and components							
	domestic hot water, air	domestic hot water, liquid	space heating, liquid	space heating, air	swimming pool	cooling	solar heat pump	combined	distillation units	process heat	direct gain glazing	greenhouses	hybrid systems	thermal storage devices	special building insulation	solar gain control	swimming pool covers	other passive systems
Lof Brothers, Solar Appliances, Inc., 1615 17th Street, Denver, CO 80202. 303/623-3410.																	•	
MacBall Industries, Inc., 1820 Embarcardero, Oakland, CA 95606. 415/534-0274.																	•	
Madico, 64 Industrial Parkway, Woburn, MA 01801. 617/935-7850.																•		
Metallized Products, 2544 Terminal Drive South, St. Petersburg, FL 33712. 813/822-9621.																•		
Mid-Western Solar & Insulation Inc., 2235 Irvin Cobb Drive, Paducah, KY 42001. 502/443-6295.		•		•			•	•										
Mor-Flo Industries, Inc., 18450 South Miles, Cleveland, OH 44128. 216/663-7300.		•			•													
Mountain Mechanical Sales, 550 East 76th Avenue, Denver, CO 80229. 800/525-1394.	•	•	•	•	•		•	•										
Multi-Research Corporation, Highway 35, Keyport, NJ 07735. 201/264-8000.	•	•	•	•	•	•		•		•								
National Metallizing, A Saxon Industries Company, P.O. Box 5202, Princeton, NJ 08540. 800/257-5116.															•	•		
Natural Energy Resources. 840 East Bethany Home, Phoenix, AZ 85014. 602/266-2949.		•			•													
Novan Energy, Inc., 3163 Walnut Street, Boulder, CO 80301. 303/447-9193.		•																

OEM Products, Inc., 2701 Adamo Drive, Tampa, FL 33605. 813/247-5947.		•	•		•		•	•		•				•				
Omnium-G, 1815.5 North Orange-Thorpe Park, Anaheim, CA 92804. 714/879-8421.										•								
One Design, Inc., Mountain Falls Route, Winchester, VA 22601. 703/662-4898.								•						•	•			
Park Energy Company, Star Route 9, Jackson, WY 83001. 307/733-4950.	•	•		•				•			•							
PerKasie Industries Corporation, 50 East Sprice Street, Perkasie, PA 18944. 215/257-6581.															•	•		
Piper Hydro, Inc., 3031 East Coronado, Anaheim, CA 92806. 714/630-4040.		•	•					•										
PolyCell Industries, Inc., P.O. Box 99, Marion, IA 52302. 319/377-9495.															•			
Porter Energy Products, P.O. Box 827, Newark, DE 19711. 301/398-0284.		•						•										
Practical Solar Systems, 7221 South 180th Street, Kent, WA 98031. 206/226-4554.	•	•	•	•										•				
Radco Products, Inc., 2877 Industrial Parkway, Santa Maria, CA 93454. 805/928-1881.		•			•													
Ra-Energy Systems, Inc., 11459 Woodside Avenue, Lakeside, CA 92040. 714/448-6216.		•	•		•			•										
Ra-Los, Inc., 559 Union Avenue, Campbell, CA 95008. 408/371-1734.		•	•					•										
Raypak, Inc., 31111 Agoura Road, Westlake Village, CA 91360. 213/889-1500.		•			•			•										
Refrigeration Research. Solar Research Division, 525 North Fifth Street, Brighton, MI 48116. 313/277-1151.		•																
Research Products Corporation, P.O. Box 1467, 1015 East Washington Avenue, Madison, WI 53713. 608/257-8801.	•	•		•				•										
Revere Solar and Architectural Products, Inc., P.O. Box 151, Rome, NY 13440. 315/338-2595.	•	•																
Rheem/Ruud Water Heater Divisions, City Investing Company, 5780 Peachtree-Dunwoody Road, Northeast, Suite 400, Atlanta, GA 30342. 404/252-7211.		•																
Richdel, Inc., P.O. Drawer A, Carson City, NV 89701. 702/882-6786.		•																
R. J. Sullivan & Sun, 1835 3rd Avenue, Southeast, Cedar Rapids, IA 52403. 319/366-0030.		•	•															
R-M Products, 5010 Cook, Denver, CO 80216. 303/825-0203.	•	•			•			•										
Rom-Aire Solar Systems, 121 Miller Road, Avon Lake, OH 44012. 216/933-5000.	•	•		•				•		•								
Sealed Air Corporation, 14428 Best Avenue, Santa Fe Springs, CA 90670. 213/921-9617.																	•	

Solar systems/passive systems and components

	Solar systems								Passive systems and components									
	domestic hot water, air	domestic hot water, liquid	space heating, liquid	space heating, air	swimming pool	cooling	solar heat pump	combined	distillation units	process heat	direct gain glazing	greenhouses	hybrid systems	thermal storage devices	special building insulation	solar gain control	swimming pool covers	other passive systems
Sebra Solar Energy, 500 North Tucson Boulevard, Tucson, AZ 85716. 602/881-6555.												•		•	•			
Shelley Radiant Ceiling Company, 456 West Frontage Road, Northfield, IL 60093. 312/446-2800.								•										
SJC Corporation, Division Frigiking-Tappan, 206 Woodford Avenue, Elyria, OH 44036. 216/329-2000.		•	•		•		•	•	•	•								
Skytherm Processes & Engineering, 2424 Wilshire Boulevard, Los Angeles, CA 90057. 213/389-2141.			•					•	•					•	•			
SMC Energy Company, P.O. Box 246, Omaha, NE 68102. 402/397-1370.		•		•				•						•				
Solafern, Ltd., 536 MacArthur Boulevard, Bourne, MA 02532. 617/563-7181.	•	•		•				•										
Solag, Rural Route 2, Roseville, IL 61473. 309/774-6641.										•		•						
Solahart, 3560 Dunhill Street, San Diego, CA 92121. 714/452-0252.		•																
Solara, Inc., 1606 Manning Boulevard, P.O. Box 245, Levittown, PA 19059. 215/757-1288.	•						•	•										
Solar Alternative, Inc., 22 South Main Street, Brattleboro, VT 05301. 802/254-6668.	•	•			•													
Solar Aquasystems, Inc., P.O. Box 88, Encintas, CA 92024. 714/753-0649.												•						

Solar by WEFCO, Inc., 324 South Kidd Street, Whitewater, WI 53190. 414/473-2525.			•	•						•								
Solar Contact Systems, 1415 Vernon, Anaheim, CA 92805. 714/991-8120.		•																
Solar Development, Inc., 3630 Reese Avenue, Riviera Beach, FL 33404. 305/842-8935.	•	•		•	•	•	•	•										
Solar Development, Inc. Northwest, 690 Yellowstone, Suite D, Pocatello, ID 83201. 208/233-6563.		•	•	•	•			•		•								
Solar Dynamics of Arizona, 1100 North Lake Havasu Avenue, Suite H, Lake Havasu City, AZ 86403. 602/855-5051.		•	•		•		•	•										
Solar-En Corporation, 3118 Route 10, Denbrook Village, Denville, NJ 07834. 201/361-2300.		•													•			
Solar Energy Components, Inc., 212 Welsh Pool Road, Lionville, PA 19353. 215/644-9017.				•							•	•		•	•	•		
Solar Energy of the South, 3266 International Drive, Mobile, AL 36606. 205/478-3647.		•			•													
Solar Energy Products, Inc., Mountain Pass, Hopewell Junction, NY 12533. 914/226-8596.		•	•		•	•		•		•								
Solar Energy Research Corporation, 1224 Sherman Drive, Longmont, CO 80501. 303/772-8406.		•	•		•	•	•	•		•								
Solar Energy Systems, Inc., P.O. Box 245, Scarsdale, NY 10583. 914/723-6353.		•	•	•	•	•				•								
Solar Enterprises, P.O. Box 1046, Red Bluff, CA 96080. 916/527-0551.					•			•										
Solar Equipment Corporation, P.O. Box 357, Lakeside, CA 92040. 714/561-4531.										•								
Solar Equipment Distributors, Inc., Division of Yago Systems Design, P.O. Box 64, Barboursville, WV 25504. 304/736-4091.		•	•				•	•										
Solar Farm Industries, Inc., P.O. Box 242, Stockton, KS 67669. 913/425-6726.	•	•		•				•										
Solar Fin Systems, 140 South Dixie Highway, St. Augustine, FL 32084. 904/824-3522.		•																
Solarflame Systems, Inc., P.O. Box 99, Leroy, IL 61752. 309/962-2861.		•	•												•			
Solargenics, Inc., 20319 Nordhoff Street, Chatsworth, CA 91311. 213/998-0806.		•	•		•	•	•	•		•								
Solargizer International, P.O. Box 20142, Minneapolis, MN 55420. 612/944-7861.		•	•	•	•		•	•										
Solar Heat Company, P.O. Box 110, Greenville, PA,	•	•	•	•				•										
Solar Heat Corporation, 398 East 271st Street, Euclid, OH 44132. 216/732-8034.	•	•	•	•	•			•										
Solar Home Systems, Inc., 8732 Camelot, Chesterland, OH 44026. 216/729-9350.		•	•	•			•	•						•	•			

	Solar systems										Passive systems and components							
	domestic hot water, air	domestic hot water, liquid	space heating, liquid	space heating, air	swimming pool	cooling	solar heat pump	combined	distillation units	process heat	direct gain glazing	greenhouses	hybrid systems	thermal storage devices	special building insulation	solar gain control	swimming pool covers	other passive systems
Solar Industries, Monmouth Airport Industrial Park, Farmingdale, NJ 07727. 201/938-7000.		•			•													
Solar Industries, Inc., Box 303, Plymouth, CT 06782. 203/283-0223.		•	•				•	•		•								
Solar Innovations, 412 Longfellow Boulevard, Lakeworth, FL 33801. 813/688-8373.		•																
Solar King International, 8577 Canoga Park, Canoga Park, CA 91304. 213/998-6400.		•	•															
Solar Living, Inc., P.O. Box 12, Netcong, NJ 07857. 201/691-8483.		•	•	•	•			•										
Solarnetics Corporation, 1654 Pioneer Way, El Cajon, CA 92020. 714/519-7122.		•			•			•										
Solarom of North America, Inc., 10 Central Avenue, Westbury, NY 11590. 516/334-1660.		•						•										
Solaron Corporation, 720 South Colorado Boulevard, Denver, CO 80222. 303/759-0101.	•	•	•	•	•		•	•		•							•	
Solar One, Ltd., 2644 Barrett Street, Virginia Beach, VA 23452. 804/340-7774.		•	•		•		•	•		•								
Solar Research Systems, 3001 Red Hill Avenue, I-105, Costa Mesa, CA 92626. 714/545-4941.					•		•	•		•								
The Solar Room Company, dba Solar Resources, Inc., Box 1848, Taos, NM 87571. 505/758-9344.											•	•		•				

Solar Servants, Inc., 289 Tropical Shore Way, Fort Myers Beach, FL 33931. 813/463-5016.		•																
Solar Specialties, Inc., Route 7, P.O. Box 409, Golden, CO 80401. 303/642-3063.		•			•			•										
Solar Spectrum, 772 W. 1355 South, Salt Lake City, UT 84104. 801/973-4207.					•													
Solar Technology Corporation, 2160 Clay, Denver, CO 80211. 303/455-3309.											•	•			•	•		
Solar Technology International, 119 North Center Street, Statesville, NC 28677. 704/873-7959.		•	•	•	•			•		•								
Solartherm, Inc., 1110 Fidler Lane, Silver Spring, MD 20910. 202/882-4000.	•	•		•	•													
Solar Thermal Systems, Division of Exxon Enterprises, Inc., 90 Cambridge Street, Burlington, MA 01803. 617/272-8460.		•	•		•	•	•	•		•								
Solar Trend Industries, Inc., P.O. Box 90633, World Way Postal Center, Los Angeles, CA 90009. 213/327-5137.		•																
Solar II Enterprises, 41 Lost Lake Lane, Campbell, CA 95008. 408/866-6244.		•	•		•		•											
Solar Unlimited, Inc., 204 Oakwood Avenue, Huntsville, AL 35811. 205/534-0661.		•	•		•		•	•										
Solar World, Inc., 4449 North 12th Street, Suite 7, Phoenix, AZ 85014. 602/266-5686.			•					•										
Solar-X Corporation, 25 Needham Street, Newton, MA 02161. 617/244-8686.															•	•		
Solcoor, Inc., 849 South Broadway·208, Los Angeles, CA 90014. 213/622-4181.		•																
Solectro-Thermo, 1934 Lakeview Avenue, Dracut, MA 01826. 617/957-0028.								•										
Solen Enterprises, P.O. Box 3355, Santa Barbara, CA 93105. 805/965-2990.		•			•													
Solenco Corporation, 175 River Road, Flanders, NJ 07836. 201/584-5055.		•																
Solex, 187 Billerica Road, Chelmsford, MA 01824. 617/256-8724.					•													
Solpower Industries, Inc., 10211-C Bubb Road, Cupertino, CA 95014. 408/996-3222.		•	•		•		•	•										
Solpub, Box 9209, College Station, TX 77840. 713/845-4133.														•	•			
Southeastern Solar Systems, Inc., 4705J Bakers Ferry Road, Atlanta, GA 30336. 404/691-1960.		•	•					•										
Space Structures International Corporation, 155 DuPont Street, Plainview, NY 11803. 516/938-0545.											•				•	•		
S.P.S. Inc., 8801 Biscayne Boulevard, Miami, FL 33138. 305/754-7766.						•												
S S Solar Inc., 16 Keystone Avenue, River Forest, IL 60305. 312/771-1912.		•	•		•	•	•			•								

Solar systems/passive systems and components

	Solar systems										Passive systems and components							
	domestic hot water, air	domestic hot water, liquid	space heating, liquid	space heating, air	swimming pool	cooling	solar heat pump	combined	distillation units	process heat	direct gain glazing	greenhouses	hybrid systems	thermal storage devices	special building insulation	solar gain control	swimming pool covers	other passive systems
Standard Solar Collectors, Inc., 1465 Gates Avenue, Brooklyn, NY 11227. 212/456-1882.	•	•		•				•		•								
Sunburst Solar Energy, Inc., P.O. Box 2799, Menlo Park, CA 94025. 415/327-8022.		•																
Sun Craft, 5001 East 59th Street, Kansas City, MO 64130. 816/333-2100.								•						•	•			
Sunduit, Inc., 281 East Jackson Street, Virden, IL 62690. 217/965-3352.		•	•		•					•								
Sunearth Solar Products Corporation, Route 1, Box 337, Green Lane, PA 18054. 215/699-7892.		•																
Sunflower Energy Works, Inc., P.O. Box 85, Goessel, KS 67053. 316/367-2647.				•														
Sun-Heet, Inc., 2624 South Zuni, Englewood, CO 80110. 303/777-2964.		•	•					•										
Sunhouse Incorporated, 28 Charron Avenue, Nashua, NH 03060. 603/889-8611.		•	•				•	•										
Sun King, 2722 West Davie Boulevard, Fort Lauderdale, FL 33312. 305/791-5415.		•																
Sunmaster Corporation, 12 Spruce Street, Corning, NY 14830. 607/937-5441.		•	•			•	•	•	•	•								
Sun-Pac, Inc., P.O. Box 8169, Alexandria, LA 71306. 318/445-6751.		•	•		•	•	•	•		•								

Company																		
Sunpower Systems Corporation, 510 South 52nd Street, Suite 101, Tempe, AZ 85281. 602/894-2331.		•	•			•		•		•								
Sun-Ray Solar Equipment Company, Inc., 415 Howe Avenue, Shelton, CT 06484. 203/735-7767.					•			•										
Sun-Ray Solar Heaters, 4898 Ronson Court, San Diego, CA 92111. 714/278-7720.		•	•		•			•										
Sunrise Solar Systems, 10 Candlelight Drive, Montvale, NJ 07645. 914/356-0032.		•			•													
Sunsav, Inc., 640 South Union Street, Lawrence, MA 01843. 617/687-0044.		•	•		•	•	•	•		•								
Sunshine Unlimited, 900 North Jay Street, Chandler, AZ 85224. 602/963-3878.		•				•		•										
Sunsource of Arizona, 3441 North 29th Avenue, Phoenix, AZ 85017. 602/258-1549.		•						•										
Sunspot Environmental Energy Systems, P.O. Box 5110, San Diego, CA 92105. 714/264-9100.	•	•	•		•		•	•										
Sun Stone Solar Energy Equipment, P.O. Box 138, Baraboo, WI 53913. 608/356-7744.	•	•		•														
Suntree Solar Company, P.O. Box 1261, Woonsocket, RI 02895. 401/769-1689.		•	•		•		•	•		•								
Sun-Wind-Home-Concepts, Division of TWR Enterprises, 72 West Meadow Lane, Sandy, UT 84070.													•	•				
Sun Wise Inc., 609 10th Avenue South, P.O. Box 6622, Great Falls, MT 59406. 406/727-5977.	•	•																
Sunworks Division, Sun Selector Corp., P.O. Box 3900, Somerville, NJ 08876. 201/469-0399.	•	•	•	•		•												
S. W. Energy Options, Route 8, Box 30-H, Silver City, NM 88061. 505/538-9598.		•		•	•			•			•	•		•	•			
Systems Technology, Inc., P.O. Box 337, Shalimar, FL 32579. 904/863-9213.		•	•		•			•										
Technitrek Corporation, 1999 Pike Avenue, San Leandro, CA 94577. 415/352-0535.		•	•		•	•		•		•								
Terra-Light, Inc., P.O. Box 493, Billerica, MA . .					•													
Texas Urethanes Inc., P.O. Box 9563, Austin, TX 78766. 512/272-5531.															•			
Thermal Technology Corporation, P.O. Box 130, Snowmass, CO 81654. 303/963-3185.															•			
Thermics, 8360A Industrial Avenue, Cotati, CA 94928. 707/526-1735.		•												•				
3-E Corporation, 401 Kennedy Boulevard, Somerdale, NJ 08083. 609/784-8200.															•			
Tri-State Sol-Aire, Inc., 631 South Jason Street, Denver, CO 80223. 303/426-4000.								•										
Turner Greenhouses, Highway 117 South, Goldsboro, NC 27530. 919/734-8345.												•						

Solar systems/passive systems and components

	Solar systems										Passive systems and components							
	domestic hot water, air	domestic hot water, liquid	space heating, liquid	space heating, air	swimming pool	cooling	solar heat pump	combined	distillation units	process heat	direct gain glazing	greenhouses	hybrid systems	thermal storage devices	special building insulation	solar gain control	swimming pool covers	other passive systems
U.S. Solar Corporation, P.O. Drawer K, Hampton, FL 32044. 904/468-1517.		•	•		•			•		•								
Valmont Energy Systems, Inc., Valley, NE 68064. 402/359-2201.	•	•		•				•										
Vanguard Energy Systems, 9133 Chesapeake Drive, San Diego, CA 92123. 714/292-1433.							•											
Vegetable Factory Inc., 100 Court Street, Copaigue, NY 11726. 516/842-9300.												•	•			•		
Virginia Solar Components, Inc., Highway 29 South, Rustburg, VA 24588. 804/239-9523.		•	•		•		•	•										
W. C. Brown & Associates, P.O. Box 5134, Portland, OR 97208. 503/289-0538.		•			•													
Weather Energy Systems, Inc., 39 Barlows Landing Road, Pocasset, MA 02559. 617/563-9337.													•					
Weather-Made Systems, Inc., Route 7, Box 300-D, Springfield, MO 65802. 417/865-0684.	•	•	•		•			•										
Western Solar Development, Inc., 1236 Callen Street, Vacaville, CA 95688. 707/446-4411.		•			•		•	•										
Westinghouse Electric, 5205 Leesburg Pike, #201, Falls Church, VA 22041. 202/833-5950.		•	•	•			•	•		•								

Wissota Solar, Inc., Rural Route 6, Box 311, Chippewa Falls, WI 54729. 715/289-3150.		•		•			•	•										
Ying Manufacturing Corporation, 1957 West 144st Street, Gardena, CA 90249. 213/770-1756.		•	•	•			•	•										
Zeopower Company, 75 Middlesex Avenue, Natick, MA 01760. 617/655-4125.						•		•										
Zomeworks Corporation, P.O. Box 712, Albuquerque, NM 87103. 505/242-5354.															•			

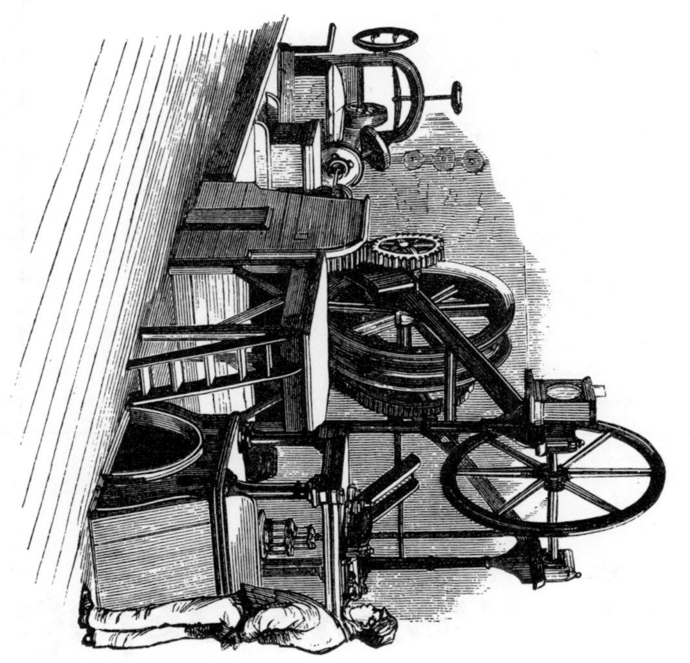

Solar system components

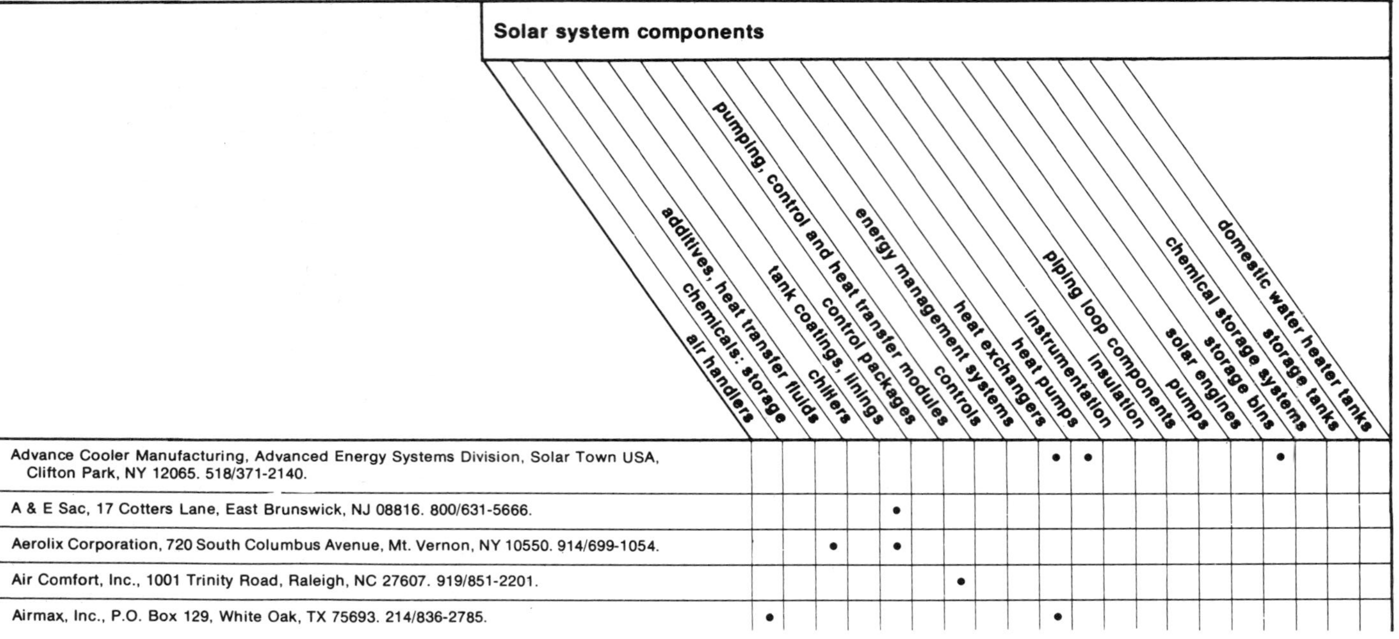

Solar system components	air handlers	chemicals: storage	additives, heat transfer fluids	chillers	tank coatings, linings	control packages	pumping, control and heat transfer modules	controls	energy management systems	heat exchangers	heat pumps	instrumentation	insulation	piping loop components	pumps	solar engines	storage bins	chemical storage systems	storage tanks	domestic water heater tanks
Advance Cooler Manufacturing, Advanced Energy Systems Division, Solar Town USA, Clifton Park, NY 12065. 518/371-2140.										•	•						•			
A & E Sac, 17 Cotters Lane, East Brunswick, NJ 08816. 800/631-5666.					•															
Aerolix Corporation, 720 South Columbus Avenue, Mt. Vernon, NY 10550. 914/699-1054.			•		•															
Air Comfort, Inc., 1001 Trinity Road, Raleigh, NC 27607. 919/851-2201.							•													
Airmax, Inc., P.O. Box 129, White Oak, TX 75693. 214/836-2785.	•									•										

All Sun Power, Inc., 10400 South West 187th, Miami, FL 33157. 305/233-2224.										•										
American Air Filter Company, 215 Central Avenue, Louisville, KY 40277. 502/637-0011.											•									
American Klegecell Corporation, 204 North Dooley Street, Grapevine, TX 76051. 817/481-3547.													•							
American Solar Products, Division of Squires Laboratories, 1996 South Valley Drive, Las Cruces, NM 88001. 505/523-5459.				•																
Ametek Power Systems Group, 1025 Polinski Road, Ivyland, PA 18974. 215/441-8770.										•										
Amprobe Instrument, 630 Merrick Road, Lynbrook, NY 11563. 516/593-5600.												•								
A. O. Smith Corporation, P.O. Box 28, Kankakee, IL 60901. 815/933-8241.										•									•	•
Aqueduct, Inc., 1934 Cotner Avenue, Los Angeles, CA 90025. 213/477-2496.														•						
Arkla Industries, P.O. Box 534, Evansville, IN 47704. 812/424-3331.	•			•																
Arklay S. Richards Company, Inc., 72 Winchester Street, Newton Highlands, MA 02161. 617/527-1512.								•				•								
Armstrong Pumps Inc., 93 East Avenue, North Tonowanda, NY 14120. 716/693-8813.														•	•					
Automation Industries, Inc., Flexible Tubing Division, P.O. Box 5698, Station B, Greenville, SC 29606. 803/288-7175.														•						
Automation Products, Inc., 3030 Max Roy Street, Houston, TX 77008. 713/869-0361.						•		•	•			•								
Bailey Instruments, 515 Victor Street, Rochelle Park, NJ 07662. 201/845-7252.								•				•								
Bard Manufacturing Company, P.O. Box 607, Bryan, OH 43506. 419/636-1194.											•									
Beeman Industries, South Road, East Hartland, CT 06024. 203/653-3073.										•										
The Bigelow Company, P.O. Box 706, New Haven, CT 06503. 203/772-3150.																			•	
Bio-Energy Systems, Inc., Box 87, Ellenville, NY 12428. 914/647-6482.										•										
Bivco Valve Corporation, 100 Production Court, New Britain, CT 06051. 203/224-9131.														•						
Blue White Industries, 14931 Chestnut Street, Westminster, CA 92683. 714/893-8529.												•								
Brown Manufacturing Company, P.O. Box 14546, Oklahoma City, OK 73114. 405/751-1323.								•												

Solar system components

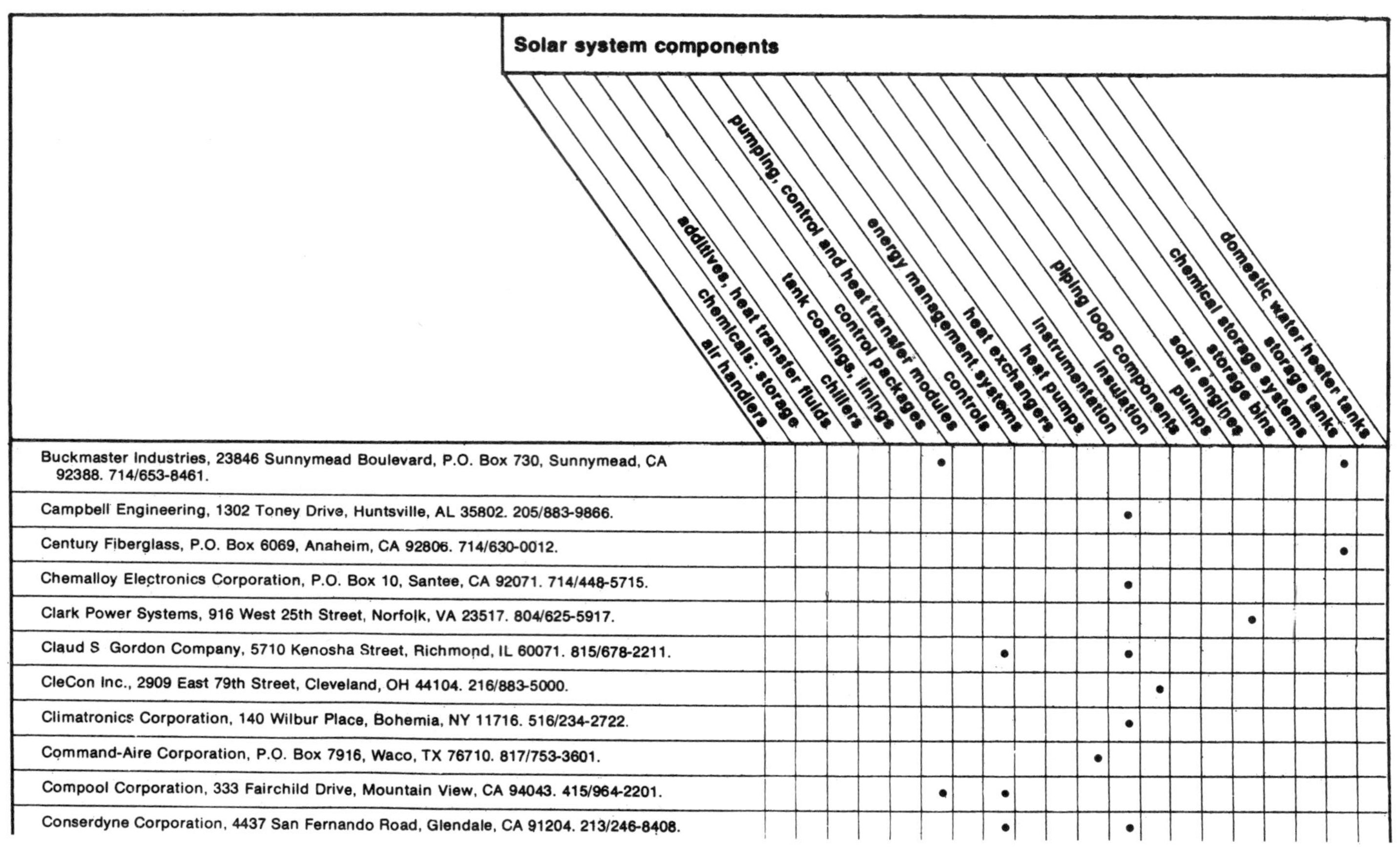

Solar system components	air handlers	chemicals: storage	additives, heat transfer fluids	chillers	tank coatings, linings	control packages	pumping, control and heat transfer modules	controls	energy management systems	heat exchangers	heat pumps	instrumentation	insulation	piping loop components	pumps	solar engines	storage bins	chemical storage systems	storage tanks	domestic water heater tanks
Buckmaster Industries, 23846 Sunnymead Boulevard, P.O. Box 730, Sunnymead, CA 92388. 714/653-8461.						•													•	
Campbell Engineering, 1302 Toney Drive, Huntsville, AL 35802. 205/883-9866.												•								
Century Fiberglass, P.O. Box 6069, Anaheim, CA 92806. 714/630-0012.																			•	
Chemalloy Electronics Corporation, P.O. Box 10, Santee, CA 92071. 714/448-5715.												•								
Clark Power Systems, 916 West 25th Street, Norfolk, VA 23517. 804/625-5917.																•				
Claud S Gordon Company, 5710 Kenosha Street, Richmond, IL 60071. 815/678-2211.								•				•								
CleCon Inc., 2909 East 79th Street, Cleveland, OH 44104. 216/883-5000.													•							
Climatronics Corporation, 140 Wilbur Place, Bohemia, NY 11716. 516/234-2722.												•								
Command-Aire Corporation, P.O. Box 7916, Waco, TX 76710. 817/753-3601.											•									
Compool Corporation, 333 Fairchild Drive, Mountain View, CA 94043. 415/964-2201.						•		•												
Conserdyne Corporation, 4437 San Fernando Road, Glendale, CA 91204. 213/246-8408.								•				•								

Company																				
Contemporary Systems, CSI Solar Center, Route 12, Walpole, NH 03603. 603/756-4796.	•							•					•				•			
Control Pak Corporation, 44480 Grand River, Nori, MI 48050. 313/349-4231.								•	•											
Copper State Solar Products, Inc., 4610 South 35th Street, P.O. Box 20504, Phoenix, AZ 85036. 602/276-4221.						•	•				•									
Cromglass Corporation, P.O. Box 3215, Williamsport, PA 17701. 717/326-3396.											•								•	
Cushing Instruments, 7911 Hershel Avenue, Suite 214, La Jolla, CA 92037. 714/459-3433.												•								
Custom Solar Heating Systems Company, P.O. Box 375, Albany, NY 12201. 518/438-7358.																	•			
The Dampney Company, 85 Paris Street, Everett, MA 02149. 617/389-2805.					•															
Datron, 20700 Plummer Street, Chatsworth, CA 91311. 213/882-9616.						•		•												
Davison Chemical Company, P.O. Box 2117, Baltimore, MD 21203. 301/659-9000.		•																		
Deko-Labs, Route 4, Box 256, Gainesville, FL 32601. 904/372-6009.								•												
Del Sol Control Corporation, 11914 U.S. Highway 1, West Palm Beach, FL 33408. 305/626-6116.						•		•												
Devices & Services Company, 3501-A Milton Avenue, Dallas, TX 75205. 214/368-5749.												•								
Dodge Products, Inc., Houston, TX,												•								
Doucette Industries, P.O. Box 1641, York, PA 17405. 717/845-8746.										•										
Dow Corning Corporation, 2200 West Salzburg Road, CO 2314, Midland, MI 48640. 517/496-4000.			•																	
Dunham-Bush, Inc., 101 Burgess Road, Harrisonburg, VA 22801. 703/434-0711.											•									
Dwyer Instruments, Inc., P.O. Box 373, Michigan City, IN 46360. 219/872-9141.								•				•								
Dynalco Corporation, Box 8187, Ft. Lauderdale, FL 33310. 305/739-4300.								•												
Dynergy Corporation, P.O. Box 428, 1269 Union Avenue, Laconia, NH 03246. 603/524-8313.												•								
Dyrelite Corporation, 63 David Street, New Bedford, MA 02741. 617/993-9955.													•							
Eaton Controls Products, 191 East North Avenue, Carol Stream, IL 60187. 312/682-8041.														•						
Eaton Metal Products Company, 4800 York Street, Denver, CO 80216. 303/825-7204.																			•	

Solar system components

Solar system components	air handlers	chemicals: storage	additives, heat transfer fluids	chillers	tank coatings, linings	control packages	pumping, control and heat transfer modules	controls	energy management systems	heat exchangers	heat pumps	instrumentation	insulation	piping loop components	pumps	solar engines	storage bins	chemical storage systems	storage tanks	domestic water heater tanks
Ecotronics, Inc., 7745 East Redfield Road, Scottsdale, AZ 85260. 602/948-8003.						•		•	•											
Electromedics, Inc., 2000 South Quebec Street, Denver, CO 80231. 303/770-8704.						•		•				•								
Elkhart Products Corporation, 1255 Oak Street, Elkhart, IN 46515. 219/264-3181.														•						
Elmwood Sensors, 1655 Elmwood Avenue, Cranston, RI 02907. 401/781-6500.								•												
Energy Alternatives, Inc., 2102 South Broadway, Wichita, KS 67211. 316/262-6483.	•					•		•									•			
Energy Control Systems, 3324 Octavia Street, Raleigh, NC 27606. 919/851-2310.	•					•		•												
Energy Distribution Inc., P.O. Box 353 Snug Harbor, Duxbury, MA 02332. 617/878-6793.																			•	
Energy Management Engineering, Inc., 4241 Hogue Road, Evansville, IN 47712. 812/464-2463.											•									
Energy Materials, Inc., 2622 South Zuni, Englewood, CO 80110. 303/934-2444.																		•		
Energy Solutions, Inc., U.S. Highway 93, P.O. Drawer J, Stevensville, MT 59870. 406/777-5640.								•	•										•	

Entropy, Ltd., 5735 Arapahoe Avenue, Boulder, CO 80303. 303/443-5103.										•										
Epic, Inc., 150 Nassau Street, New York, NY 10038. 212/349-2470.												•								
The Eppley Laboratory, Inc., 12 Sheffield Avenue, Newport, RI 02840. 401/847-1020.												•								
Fabrico Manufacturing Corporation, 1300 West Exchange Avenue, Chicago, IL 60609. 312/254-4211.					•															
Ferro Corporation, One Erie View Plaza, Cleveland, OH 44114. 216/641-8580.					•															
Fiber-Rite Products, Inc., P.O. Box 38095, Cleveland, OH 44138. 216/235-6800.					•														•	
Flexonics Division, UOP, Inc., 300 East Devon Avenue, Bartlett, IL 60103. 312/837-1811.										•				•						
Ford Products Corporation, Ford Products Road, Valley Cottage, NY 10989. 914/358-8282.																				•
F. W. Bell, Inc., 6120 Hanging Moss Road, Orlando, FL 32807. 305/678-6900.												•								
Gamma Scientific Inc., 3777 Ruffin Road, San Diego, CA 92123. 714/279-8034.												•								
GCA/Precision Scientific Group, 3737 West Cortland Street, Chicago, IL 60647. 312/227-2660.												•			•					
Gem Manufacturing Corporation, 390 East Tiffin Street, Bascom, OH 44809. 419/937-2225.								•				•							•	
General Energy Devices, Inc., 1753 Ensley Avenue, Clearwater, FL 33516. 813/536-3585.										•									•	
General Scientific Equipment Company, Limekiln Pike and Williams Avenue, Philadelphia, PA 19150. 215/424-1550.														•	•					
The Glass-Lined Water Heater Company, 13000 Athens Avenue, Lakewood, OH 44107. 216/521-1377.										•									•	•
Gould Pumps, 240 Fall Street, Seneca Falls, NY 13148. 315/568-2811.															•					
Grumman Energy Systems, Inc., 4175 Veterans Memorial Highway, Ronkonkoma, NY 11779. 516/737-3777.										•									•	
Grundfos Pumps Corporation, 2555 Clovis Avenue, Clovis, CA 93612. 209/299-9741.															•					
Gulton MCS Division, East Greenwich, RI 02818. 401/884-6800.												•								
Halstead & Mitchell, P.O. Box 1110, Scottsboro, AL 35768. 205/259-1212.										•										
Hawthorne Industries, Inc., 1501 South Dixie Highway, West Palm Beach, FL 33401. 305/659-5400.						•		•				•								

Solar system components

Solar system components	air handlers	chemicals: storage	additives, heat transfer fluids	chillers	tank coatings, linings	control packages	pumping, control and heat transfer modules	controls	energy management systems	heat exchangers	heat pumps	instrumentation	insulation	piping loop components	pumps	solar engines	storage bins	chemical storage systems	storage tanks	domestic water heater tanks
Heat Controller, Inc., 1900 Wellworth Avenue, Jackson, MI 49203. 517/787-2100.												•								
Heliodyne, Inc., 770 South 16th Street, Richmond, CA 94804. 415/237-9614.							•													
Heliotrope General, 3733 Kenora Drive, Spring Valley, CA 92077. 714/460-3930.								•												
H & H Precision Products, Emerson Electric Company, 25 Canfield Road, Cedar Grove, NJ 07009. 201/239-1331.															•				•	
Hi-Tech Inc., 3600 16th Street, Zion, IL 60099. 312/746-2447.																•				
Hollis Observatory, One Pine Street, Nashua, NH 03060. 603/882-5017.													•							
Hometech, 266 Viking Avenue, Brea, CA 92621. 714/990-5131.								•	•				•							
Honeywell Inc., Honeywell Plaza, Minneapolis, MN 55408.						•		•	•											
Hubbell, The Electric Heater Company, 45 Seymour Street, Stratford, CT 06497. 203/378-2659.																			•	•
Hydra-Air Equipment Company, 517 West Garfield Street, Glendale, CA 91204. 213/246-8276.	•																			
Hydrocap Corporation, 3579 East 10th Court, Hialeah, FL 33013. 305/696-2504.						•							•							

Company																				
Hydro-Flex Corporation, 2101 N.W. Brickyard Road, Topeka, KS 66618. 913/233-7484.										•				•						
Hyperion, Inc., 4860 Riverbend Road, Boulder, CO 80303. 303/449-9544.							•			•										
ILI, Inc., 5965 Peachtree Corners East, Atlanta, GA 30071. 404/449-5900.						•	•			•		•							•	
IMC Instruments, Inc., 6659 North Sidney Place, Glendale, WI 53209. 414/352-3810.												•								
Impac Corporation, 312 Blondeau Street, Keokuk, IA 52632. 319/524-3304.	•																			
Independent Energy, Inc., Box 732, 42 Ladd Street, East Grennwich, RI 02818. 401/884-6990.						•		•	•			•								
Insultek Corporation, 82 Crestwood Road, Rockaway, NJ 07866.													•							
Intec, 1131 Lincoln Highway, North Versailles, PA 15137. 412/824-8100.								•							•					
International Technology Corporation, 1670 Highway AIA, Satellite Beach, FL 32937. 305/777-1400.									•			•								
International Thermal Instrument Company, P.O. Box 309, Del Mar, CA 92014. 714/755-4436.												•								
Isophenol Division, Robert Mitchell Solar Systems Design, Route 3, Box 239, Selkirk, NY 12158. 518/767-3100.													•							
ITT, Fluid Handling Division, 4711 Golf Road, Skokie, IL 60076.							•			•					•				•	
John M. Frey Company, P.O. Box 560, 530 34th Street, Moline, IL 61265. 309/797-1161.														•	•					
Johnson Controls, Inc.,, Control Products Division, 2221 Camden Court, Oak Brook, IL 60521. 312/654-4900.								•												
Kalwall Corporation, P.O. Box 237, Manchester, NH 03105. 603/627-3861.								•											•	
Kipp & Zonen, 390 Central Avenue, Bohemia, NY 11716. 516/589-2885.												•								
KNF Neuberger, Inc., P.O. Box 4060, Princeton, NJ 08540. 609/799-4350.															•					
Koldwave Division, Heat Exchangers, Inc., 8100 North Monticello, Skokie, IL 60076. 312/267-8282.											•									
Lafont Corporation, 1319 Town Street, Prentice, WI 54556. 715/428-2881.										•										
Largo Solar Systems, Inc., 991 South 40th Avenue, Ft. Lauderdale, FL 33317. 305/583-8090.										•										

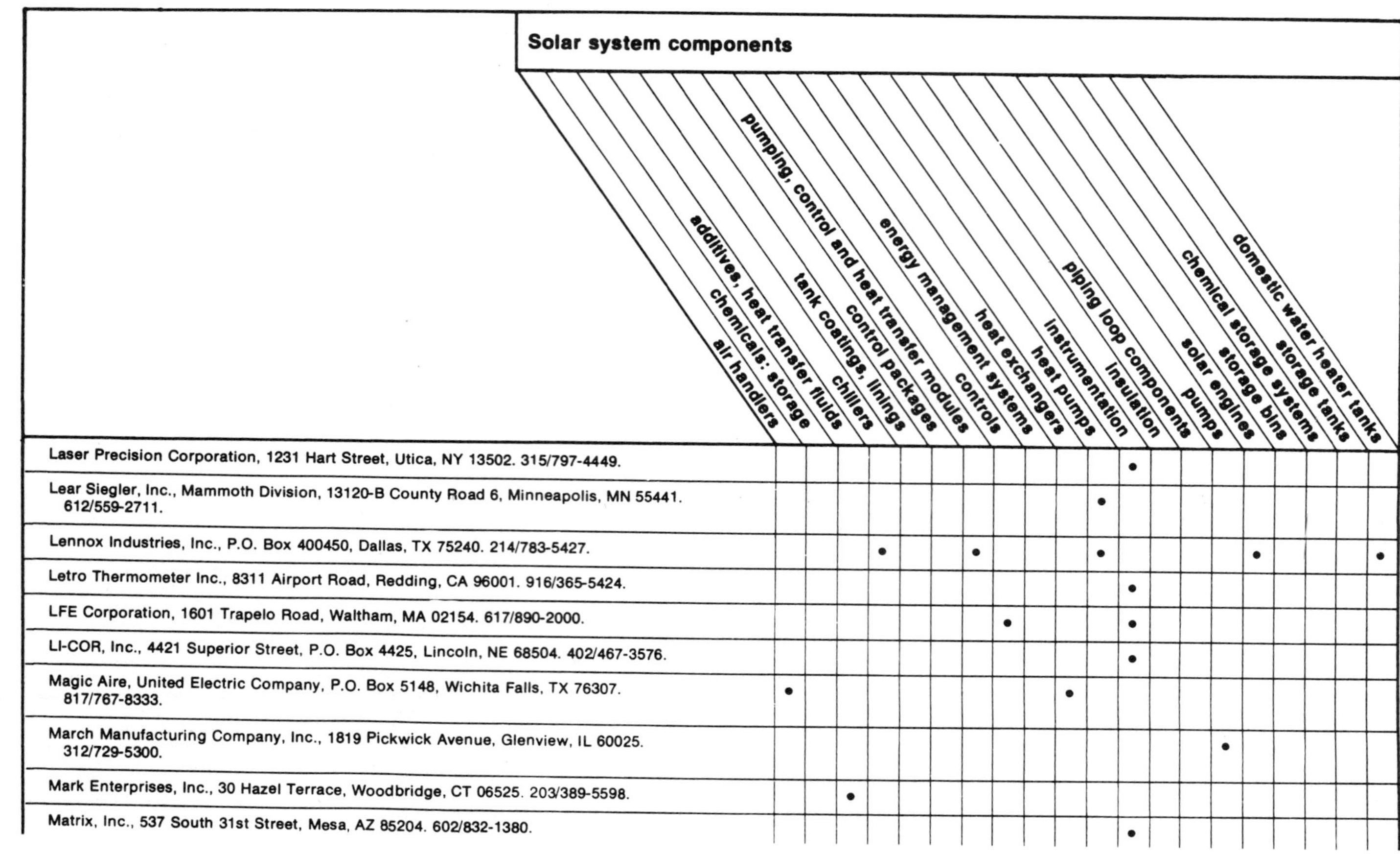

Solar system components	air handlers	chemicals: storage	additives, heat transfer fluids	chillers	tank coatings, linings	control packages	pumping, control and heat transfer modules	controls	energy management systems	heat exchangers	heat pumps	instrumentation	insulation	piping loop components	pumps	solar engines	storage bins	chemical storage systems	storage tanks	domestic water heater tanks
Laser Precision Corporation, 1231 Hart Street, Utica, NY 13502. 315/797-4449.												•								
Lear Siegler, Inc., Mammoth Division, 13120-B County Road 6, Minneapolis, MN 55441. 612/559-2711.											•									
Lennox Industries, Inc., P.O. Box 400450, Dallas, TX 75240. 214/783-5427.				•			•				•					•				•
Letro Thermometer Inc., 8311 Airport Road, Redding, CA 96001. 916/365-5424.												•								
LFE Corporation, 1601 Trapelo Road, Waltham, MA 02154. 617/890-2000.								•				•								
LI-COR, Inc., 4421 Superior Street, P.O. Box 4425, Lincoln, NE 68504. 402/467-3576.												•								
Magic Aire, United Electric Company, P.O. Box 5148, Wichita Falls, TX 76307. 817/767-8333.	•									•										
March Manufacturing Company, Inc., 1819 Pickwick Avenue, Glenview, IL 60025. 312/729-5300.															•					
Mark Enterprises, Inc., 30 Hazel Terrace, Woodbridge, CT 06525. 203/389-5598.			•																	
Matrix, Inc., 537 South 31st Street, Mesa, AZ 85204. 602/832-1380.												•								

Meteorology Research, Inc., 464 West Woodbury Road, Altadena, CA 91001. 213/791-1901.											●									
The Metraflex Company, 2323 West Hubbard Street, Chicago, IL 60612. 312/738-3800.													●							
MicroControl Systems, Inc., 6579 North Sidney Place, Milwaukee, WI 53209. 414/351-0281.								●												
Midwest Components, Inc., Port City Industrial Park, Box 787, Muskegon, MI 49443. 616/777-2602.							●				●									
Milton Roy Company, Hartell Division, 70 Industrial Drive, Ivyland, PA 18974. 215/322-0730.														●						
Minco Products, Inc., 7300 Commerce Lane, Minneapolis, MN 55432. 612/571-3121.											●									
MNK Enterprises, Inc., P.O. Box 87, Bancroft, ID 83217. 208/648-7668.																		●		
Monitor Labs, 10180 Scripps Ranch Boulevard, San Diego, CA 92131. 714/578-5060.								●			●									
Mor-Flo Industries, Inc., 18450 South Miles, Cleveland, OH 44128. 216/663-7300.									●				●					●	●	
Mueller Brass Company, 1925 Lapeer Avenue, Port Huron, MI 48060. 313/987-4000.													●							
Multi-Duti Manufacturing Company, 9660 East Rush Street, South El Monte, CA 91733. 213/443-1784.														●						
Multiform Dessiccant Products, Inc., 1418 Niagara Street, Buffalo, NY 14213. 716/883-8900.		●																		
Myson, Inc., P.O. Box 5025, Embrey Industrial Park, Falmonth, VA 22401. 703/371-4331.														●						
Natural Power, Inc., New Boston, NH 03070. 603/487-5512.							●				●									
OEM Products, Inc., 2701 Adamo Drive, Tampa, FL 33605. 813/247-5947.		●							●	●								●		
Ogontz Controls Company, P.O. Box 479, Willow Grove, PA 19090. 215/657-4770.													●							
Omega Engineering, Inc., 1 Omega Drive, Largo Industrial Park, Stamford, CT 06907. 203/359-1660.							●				●									
Omnium-G, 1815.5 North Orange-Thorpe Park, Anaheim, CA 92804. 714/879-8421.															●					
One Design, Inc., Mountain Falls Route, Winchester, VA 22601. 703/662-4898.																		●		
Oven Industries, P.O. Box 229, Hempt Road, Mechanicsburg, PA 17055. 717/766-0721.							●													

Solar system components

Company	air handlers	chemicals: storage	additives, heat transfer fluids	chillers	tank coatings, linings	control packages	pumping, control and heat transfer modules	controls	energy management systems	heat exchangers	heat pumps	instrumentation	insulation	piping loop components	pumps	solar engines	storage bins	chemical storage systems	storage tanks	domestic water heater tanks
Pacific Fiberglass Company, 1015 East Elm Avenue, Fullerton, CA 92531. 714/992-1890.													•							
P.A.C.O. Pacific Pumps, Division of Baltimore Aircoil, 845 92nd Avenue, Oakland, CA 94604. 415/562-5628.															•					
Pak-Tronics, Inc., 4044 North Rockwell, Chicago, IL 60618. .						•		•				•								
Park Energy Company, Star Route 9, Jackson, WY 83001. 307/733-4950.	•																			
Permaloy Corporation, P.O. Box 1559, Ogden, UT 84402. 801/731-4303.	•		•														•			
Phelps Dodge Brass Company, Solar Division, 1050 Wall Street West, Lyndhurst, NJ 07071. 201/935-8180.														•						
Pittsburgh Corning Corporation, 800 Presque Isle Drive, Pittsburgh, PA 15239. 412/327-6100.													•							
PolyCell Industries, Inc., P.O. Box 99, Marion, IA 52302. 319/377-9495.													•							
Porter Energy Products, P.O. Box 827, Newark, DE 19711. 301/398-0284.																			•	
Premier Pump and Pool Products, Inc., 3347 San Fernando Road, Los Angeles, CA 90065. 213/258-8000.															•				•	

PSI Energy Systems, Inc., 1533 Fenpark Drive, St. Louis, MO 63026. 314/343-7666.																		•		
Pyco, Inc., 600 East Lincoln Highway, Langhorne, PA 19047. 215/757-3704.								•				•								
Pyramid Controls, P.O. Box 2211, Martinez, CA 94553. 415/229-2470.						•		•				•								
Rainaire Products, 1300 Admiral Wilson Boulevard, Camden, NJ 08109. 609/966-8500.														•						
Ra-Los, Inc., 559 Union Avenue, Campbell, CA 95008. 408/371-1734.								•												
Refrigeration Research, Solar Research Division, 525 North Fifth Street, Brighton, MI 48116. 313/277-1151.										•										
Refrigeration Systems Company, 4241 Hogue Road, Evansville, IN 47712. 812/464-9592.				•						•	•									
Revere Solar and Architectural Products, Inc., P.O. Box 151, Rome, NY 13440. 315/338-2595.							•													
Rheem/Ruud Water Heater Divisions, City Investing Company, 5780 Peachtree-Dunwoody Road, Northeast, Suite 400, Atlanta, GA 30342. 404/252-7211.																			•	•
Rho Sigma, 11922 Valerio Street, North Hollywood, CA 91605. 213/982-6800.								•				•								
Richdel, Inc., P.O. Drawer A, Carson City, NV 89701. 702/882-6786.						•	•	•						•	•					
R. J. Sullivan & Sun, 1835 3rd Avenue, Southeast, Cedar Rapids, IA 52403. 319/366-0030.								•												
R-M Products, 5010 Cook, Denver, CO 80216. 303/825-0203.	•																			
R. M. Young Company, 2801 Aero Park Drive, Traverse City, MI 49684. 616/946-3980.												•								
Robertshaw Controls Company, 100 West Victoria Street, Long Beach, CA 90805. 213/638-6111.								•												
Rotoflow Corporation, 2235 Carmelina Avenue, Los Angeles, CA 90064. 213/477-3083.																•				
Rovanco Corp., Interstate 55 & Frontage Road, Joliet, IL 60436. 800/435-0145.													•							
Rox International Inc., 2604D Hidden Lake Drive North, Sarasota, FL 33577. .																•				
Safety Direct Inc., 23 Snider Way, Sparks, NV 89431. 702/359-4451.												•								
Sherwood Specialties, Inc., 412 Smith Street, Rochester, NY 14608. 716/454-3970.															•					
SHURFlo, 1400 Cerritos Avenue East, Anaheim, CA 92805. 800/854-3218.															•					

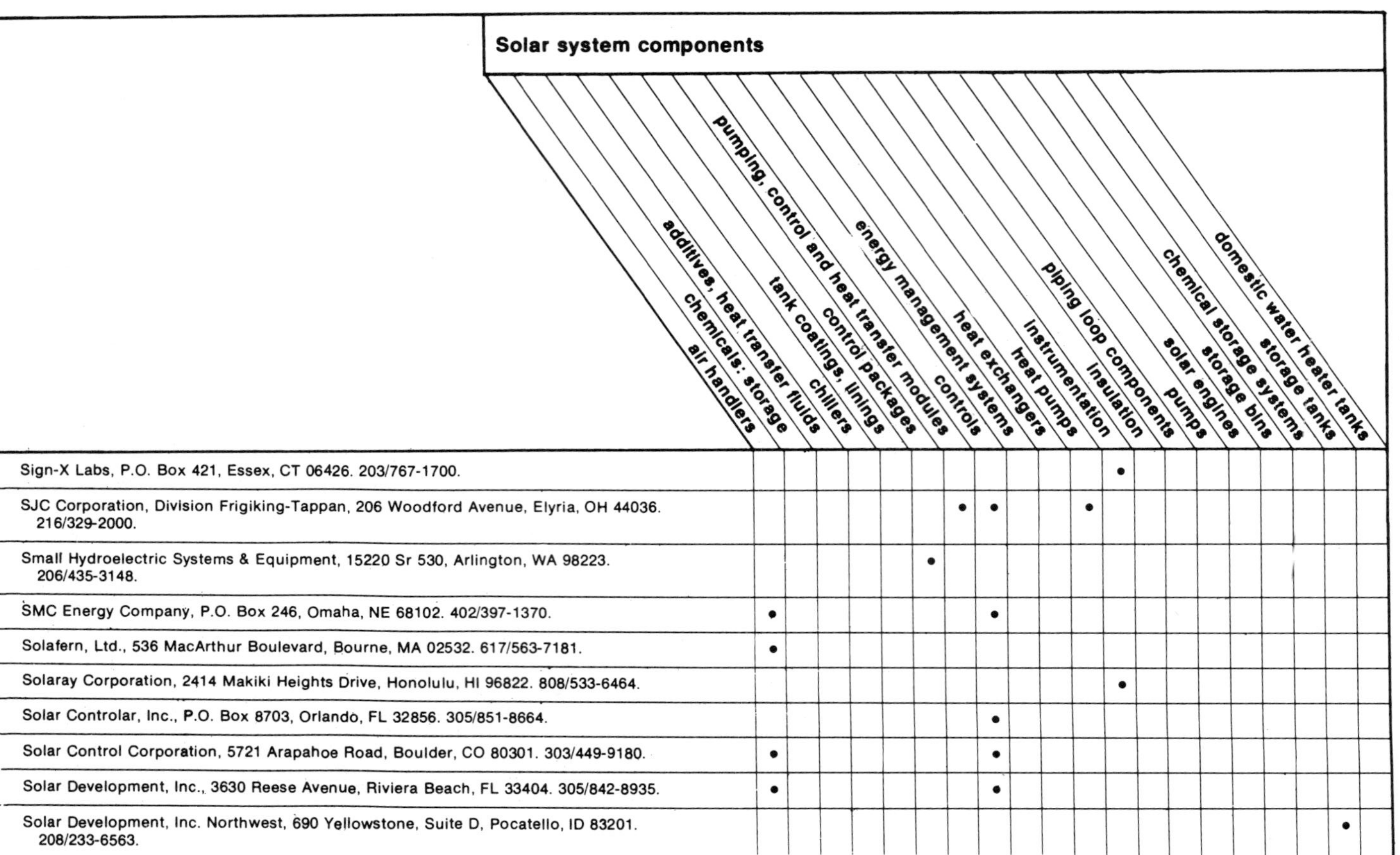

Solar system components	air handlers	chemicals: storage	additives, heat transfer fluids	chillers	tank coatings, linings	control packages	pumping, control and heat transfer modules	controls	energy management systems	heat exchangers	heat pumps	instrumentation	insulation	piping loop components	pumps	solar engines	storage bins	chemical storage systems	storage tanks	domestic water heater tanks
Sign-X Labs, P.O. Box 421, Essex, CT 06426. 203/767-1700.												•								
SJC Corporation, Division Frigiking-Tappan, 206 Woodford Avenue, Elyria, OH 44036. 216/329-2000.							•	•			•									
Small Hydroelectric Systems & Equipment, 15220 Sr 530, Arlington, WA 98223. 206/435-3148.						•														
SMC Energy Company, P.O. Box 246, Omaha, NE 68102. 402/397-1370.	•							•												
Solafern, Ltd., 536 MacArthur Boulevard, Bourne, MA 02532. 617/563-7181.	•																			
Solaray Corporation, 2414 Makiki Heights Drive, Honolulu, HI 96822. 808/533-6464.												•								
Solar Controlar, Inc., P.O. Box 8703, Orlando, FL 32856. 305/851-8664.								•												
Solar Control Corporation, 5721 Arapahoe Road, Boulder, CO 80301. 303/449-9180.	•							•												
Solar Development, Inc., 3630 Reese Avenue, Riviera Beach, FL 33404. 305/842-8935.	•							•												
Solar Development, Inc. Northwest, 690 Yellowstone, Suite D, Pocatello, ID 83201. 208/233-6563.																			•	

Solar Dynamics of Arizona, 1100 North Lake Havasu Avenue, Suite H, Lake Havasu City, AZ 86403. 602/855-5051.						•		•												
Solar Energy Corporation, 553 Pretty Brook Road, Princeton, NJ 08540. 609/924-1879.												•								
Solar Energy Engineering, 31 Maxwell Court, Santa Rosa, CA 95401. 707/542-4498.	•																			
Solar Energy Products, Inc., Mountain Pass, Hopewell Junction, NY 12533. 914/226-8596.							•	•												
Solar Energy Research Corporation, 1224 Sherman Drive, Longmont, CO 80501. 303/772-8406.				•						•	•								•	•
Solar Engines, 2937 West Indian School Road, Phoenix, AZ 85017. 602/274-3541.																•				
Solar Enterprises, P.O. Box 1046, Red Bluff, CA 96080. 916/527-0551.										•										
Solar Enterprises, Inc., 7928 N.E. Main Street, Fridley, MN 55432. 612/784-7177.								•		•	•			•	•				•	
Solar Environmental Engineering, 2524 East Vine Drive, Ft. Collins, CO 80524. 303/221-5166.								•												
Solar Equipment Corporation, P.O. Box 357, Lakeside, CA 92040. 714/561-4531.								•								•				
Solar Equipment Distributors, Inc., Division of Yago Systems Design, P.O. Box 64, Barboursville, WV 25504. 304/736-4091.								•												
Solar-Eye Products, Inc., 1300 N.W. McNab, Fort Lauderdale, FL 33309. 305/974-2500.						•														
Solar Farm Industries, Inc., P.O. Box 242, Stockton, KS 67669. 913/425-6726.	•																			
Solargenics, Inc., 20319 Nordhoff Street, Chatsworth, CA 91311. 213/998-0806.	•						•	•		•										
Solar Heat Company, P.O. Box 110, Greenville, PA,	•									•							•			
Solar Industries, Inc., Box 303, Plymouth, CT 06782. 203/283-0223.																			•	
Solar Innovations, 412 Longfellow Boulevard, Lakeworth, FL 33801. 813/688-8373.								•		•										
Solar King International, 8577 Canoga Park, Canoga Park, CA 91304. 213/998-6400.						•		•												
Solar Supply, Inc., 6709 Convoy Court, San Diego, CA 92123. 714/292-7811.						•	•	•												
Solar Technology International, 119 North Center Street, Statesville, NC 28677. 704/873-7959.										•										
Solar Thermal Systems, Division of Exxon Enterprises, Inc., 90 Cambridge Street, Burlington, MA 01803. 617/272-8460.			•																	

Solar system components

Company	air handlers	chemicals: storage	additives, heat transfer fluids	chillers	tank coatings, linings	control packages	pumping, control and heat transfer modules	controls	energy management systems	heat exchangers	heat pumps	instrumentation	insulation	piping loop components	pumps	solar engines	storage bins	chemical storage systems	storage tanks	domestic water heater tanks
Solar Unlimited, Inc., 204 Oakwood Avenue, Huntsville, AL 35811. 205/534-0661.						•	•	•		•									•	
Solar World, Inc., 4449 North 12th Street, Suite 7, Phoenix, AZ 85014. 602/266-5686.						•	•	•	•			•								
Solcoor, Inc., 849 South Broadway 208, Los Angeles, CA 90014. 213/622-4181.										•										
Solecon, P.O. Box 09763, 2770 East Main Street, Columbus, OH 43209. 614/236-5982.		•																		
Solid State Solar Controls, 123 Independence Drive, Menlo Park, CA 94025. 415/324-1980.						•		•												
Southeastern Solar Systems, Inc., 4705J Bakers Ferry Road, Atlanta, GA 30336. 404/691-1960.										•									•	
Spectran/Instruments, Inc., P.O. Box 891, La Habra, CA 90631. 213/694-3995.												•								
S.P.S. Inc., 8801 Biscayne Boulevard, Miami, FL 33138. 305/754-7766.																	•			
Standard Solar Collectors, Inc., 1465 Gates Avenue, Brooklyn, NY 11227. 212/456-1882.																				•
Sun Craft, 5001 East 59th Street, Kansas City, MO 64130. 816/333-2100.	•																			
Sun-Heet, Inc., 2624 South Zuni, Englewood, CO 80110. 303/777-2964.																			•	

Sunhouse Incorporated, 28 Charron Avenue, Nashua, NH 03060. 603/889-8611.									•	•									•	
Sun King, 2722 West Davie Boulevard, Fort Lauderdale, FL 33312. 305/791-5415.							•													
Sunpower Systems Corporation, 510 South 52nd Street, Suite 101, Tempe, AZ 85281. 602/894-2331.								•												
Sun-Ray Solar Equipment Company, Inc., 415 Howe Avenue, Shelton, CT 06484. 203/735-7767.								•												
Sunshine Power Company, 1018 Lancer Drive, San Jose CA 95129. 408/446-2446.						•		•	•			•								
Sunsource of Arizona, 3441 North 29th Avenue, Phoenix, AZ 85017. 602/258-1549.							•													
Sunspool Corporation, 439 Tasso Street, Palo Alto, CA 94301. 415/328-4718.														•						
Sunspot Division, Elcam, Inc., 5330 Debbie Lane, Santa Barbara, CA 93111. 805/964-8676.								•												
Sun Stone Solar Energy Equipment, P.O. Box 138, Baraboo, WI 53913. 608/356-7744.	•																			
Sun Systems, Inc., P.O. Box 347, Boston, MA 02186. 617/265-9600.						•						•								
Sunverter Company, Inc., Route 1, Box 269, Murphysboro, IL 62966. 618/687-3416.	•					•		•	•								•	•		
Sun Wise Inc., 609 10th Avenue South, P.O. Box 6622, Great Falls, MT 59406. 406/727-5977.								•									•			
Sunworks Division, Sun Solector Corp., P.O. Box 3900, Somerville, NJ 08876. 201/469-0399.			•																	
Surgeonics, Ltd., 155 Kisco Avenue, Mt. Kisco, NY 10549. 914/428-1954.									•											
S. W. Energy Options, Route 8, Box 30-H, Silver City, NM 88061. 505/538-9598.														•						
Syscon International, Inc., 1239 South Bend Avenue, South Bend, IN 46617. 219/287-5916.						•		•	•			•								
Systems Technology, Inc., P.O. Box 337, Shalimar, FL 32579. 904/863-9213.							•													
TACO, 1160 Cranston Street, Cranston, RI 02920. 401/942-8000.							•								•					
Texas Controls, Inc., 13735 Omega, Dallas, TX 75229. 214/386-5000.									•			•								
Texas Electronics, Inc., 5529 Redfield Street, Dallas, TX 75209. 214/631-2490.												•								
Texas Urethanes Inc., P.O. Box 9563, Austin, TX 78766. 512/272-5531.													•							

Solar system components

Solar system components	air handlers	chemicals: storage	additives, heat transfer fluids	chillers	tank coatings, linings	control packages	pumping, control and heat transfer modules	controls	energy management systems	heat exchangers	heat pumps	instrumentation	insulation	piping loop components	pumps	solar engines	storage bins	chemical storage systems	storage tanks	domestic water heater tanks
Thermics, 8360A Industrial Avenue, Cotati, CA 94928. 707/526-1735.													•						•	•
3-E Corporation, 401 Kennedy Boulevard, Somerdale, NJ 08083. 609/784-8200.					•															
Thrush Products, Inc., P.O. Box 228, Peru, IN 46970. 317/472-3351.										•				•	•				•	
Trane Company, 3600 Pammel Creek Road, La Crosse, WI 54601. 608/787-3111.	•			•						•	•									
T S I Inc., P.O. Box 43394, St. Paul, MN 55164. 612/483-0900.												•								
Turner Greenhouses, Highway 117 South, Goldsboro, NC 27530. 919/734-8345.																			•	
Ultra-Violet Products, Inc., 5100 Walnut Grove Avenue, San Gabriel, CA 91778. 213/285-3123.												•								
Universal Enterprises, Inc., 14270 Northwest Science Park Drive, Portland, OR 97229. 503/644-8723.												•								
Urethane Molding, Inc., R.F.D. 3, Route 11, Laconia, NH 03246. 603/524-7577.													•							
U.S. Solar Corporation, P.O. Drawer K, Hampton, FL 32044. 904/468-1517.								•											•	•

Valley Forge Instrument Company, Inc., 55 Buckwalter Road, Phoenixville, PA 19460. 215/933-1806.								●												
Vanguard Energy Systems, 9133 Chesapeake Drive, San Diego, CA 92123. 714/292-1433.											●									
Vaughn Corporation, 386 Elm Street, Newburyport, MA 01950. 617/462-6683.																				●
Virginia Solar Components, Inc., Highway 29 South, Rustburg, VA 24588. 804/239-9523.							●							●						
Weather Energy Systems, Inc., 39 Barlows Landing Road, Pocasset, MA 02559. 617/563-9337.						●		●				●								
Weathertronics, Inc., 2777 Del Monte Street, West Sacramento, CA 95691. 916/371-2660.												●								
Weksler Instruments, 80 Mill Road, Freeport, NY 11520. 516/623-0100.												●								
Wendland Manufacturing Company, Box 808, San Angelo, TX 76902. 915/655-6778.																			●	
Westberg Manufacturing, Inc., 3400 Westach Way, Sonoma, CA 95476. 707/938-2121.												●								
West Instruments, Gulton MCS Division, East Greenwich, RI 02818. 401/884-6800.								●				●								
West Wind Electronics, Inc., P.O. Box 542, Durango, CO 81301. 303/247-9677.								●												
William Printz Company, 81 Urban Avenue, Westbury, NY 11590. 516/334-7770.					●															
Willtronix, 1927 Clifton Avenue, Royal Oak, MI 48073. 313/399-9557.								●												
Windworks, Inc., Box 44A, Route 3, Mukwonago, WI 53149. 414/363-4088.								●	●											
Xencon, 150 Mitchell Boulevard, San Rafael, CA 94903. 415/472-5540.						●														
Ying Manufacturing Corporation, 1957 West 144st Street, Gardena, CA 90249. 213/770-1756.	●																			
Zeopower Company, 75 Middlesex Avenue, Natick, MA 01760. 617/655-4125.				●																
Zia Associates, Inc., 5590 Arapahoe, P.O. Box 1466, Boulder, CO 80306. 303/449-9170.						●		●	●											

Power generation
Solar-related products

	Power generation systems and components								Other related systems and products					
	photovoltaics	solar thermal power generation	small hydraulic generators	ocean thermal energy conversion	space power satellite	wind generation systems	wind turbine generators	other wind generation components	electrical storage	heat recovery systems, components	biomass	solar tools and equipment	design aids, educational packages	factory-built homes
A A I Corporation, P.O. Box 6767, Baltimore, MD 21204. 301/666-1400.	•	•			•	•		•		•	•	•		
AFG Industries, P.O. Box 929, Kingsport, TN 37662. 615/245-0211.	•													
Alpha Solarco, 1014 Vine Street, Suite 2230, Cincinnati, OH 45202. 513/621-1243.		•												
American Solar Products, Division of Squires Laboratories, 1996 South Valley Drive, Las Cruces, NM 88001. 505/523-5459.	•													
ARCO Solar, Inc., 20554 Plummer Street, Chatsworth, CA 91311. 213/998-0667.	•													

Arkla Industries, P.O. Box 534, Evansville, IN 47704. 812/424-3331.										•		•		
Aton Solar Manufacturing, 20 Pamaron, Novato, CA 94947. 415/883-0866.		•												
ATR Electronics, Inc., 300 East 4th Street, St. Paul, MN 55101. 612/222-3791.								•						
Automatic Power, Inc., 213 Hutcheson Street, Houston, TX 77003. 713/228-5208.	•					•	•							
Ballard Concrete Company, Inc., P.O. Box 7175, Greenville, SC 29610. 803/295-0610.										•				
Beeman Industries, South Road, East Hartland, CT 06024. 203/653-3073.												•		
Best Energy Systems for Tomorrow, Route 1, P.O. Box 106, Necedah, WI 54646. 608/565-7200.	•					•		•						
The Bigelow Company, P.O. Box 706, New Haven, CT 06503. 203/772-3150.										•				
Bio-Gas of Colorado, Inc., 5620 Kendall Court, Unit G, Arvada, CO 80002. 303/422-4354.											•			
Bio Solar Research and Development Corporation, 1500 Valley River Drive, Eugene, OR 97401. 503/686-0765.											•			
Briquettor Systems, Inc., P.O. Box 477, Reedsport, OR 97467. 503/271-2625.											•			
Carter Motor Company, 2745 George Street, Chicago, IL 60618. 312/588-7700.								•						
Chalk Wind Systems, P.O. Box 446, St. Cloud, FL 32713. 305/892-7338.							•							
Clark Power Systems, 916 West 25th Street, Norfolk, VA 23517. 804/625-5917.		•												
Daniel Enterprises, Inc., P.O. Box 2370, La Habra, CA 90631. 213/943-8883.													•	
Dow Corning Corporation, 2200 West Salzburg Road, CO 2314, Midland, MI 48640. 517/496-4000.	•													
Dynergy Corporation, P.O. Box 428, 1269 Union Avenue, Laconia, NH 03246. 603/524-8313.						•	•	•						
Ecotronics, Inc., 7745 East Redfield Road, Scottsdale, AZ 85260. 602/948-8003.	•													
Electrolab, Inc., 2103 Mannix, San Antonio, TX 78217. 512/824-5364.	•													
Energy Design Corporation, P.O. Box 34294, Memphis, TN 38134. 901/382-3000.		•												
Energy Distribution Inc., P.O. Box 353 Snug Harbor, Duxbury, MA 02332. 617/878-6793.										•				
Energy Solutions, Inc., U.S. Highway 93, P.O. Drawer J, Stevensville, MT 59870. 406/777-5640.													•	
Enertech, Box 420, Norwich, VT 05055. 802/649-1145.						•	•							
Entropy, Ltd., 5735 Arapahoe Avenue, Boulder, CO 80303. 303/443-5103.										•	•	•	•	

Power generation/solar-related products

	Power generation systems and components									Other related systems and products				
	photovoltaics	solar thermal power generation	small hydraulic generators	ocean thermal energy conversion	space power satellite	wind generation systems	wind turbine generators	other wind generation components	electrical storage	heat recovery systems, components	biomass	solar tools and equipment	design aids, educational packages	factory-built homes
ERGENICS, 681 Lawlins Road, Wyckoff, NJ 07481. 201/891-9103.	•									•				
Flanagan's Plans, P.O. Box 891, Cathedral Station, New York, NY 10025. 212/222-4774.							•	•						
Forest Fuels Manufacturing, Inc., Clinton Road, Antrim, NH 03440. 603/357-3311.											•			
F. W. Bell, Inc., 6120 Hanging Moss Road, Orlando, FL 32807. 305/678-6900.								•						
General Electric Company, P.O. Box 13601, Philadelphia, PA 19101. 215/962-2112.	•						•							
Grumman Energy Systems, Inc., 4175 Veterans Memorial Highway, Ronkonkoma, NY 11779. 516/737-3777.							•	•					•	
Horizon Industries, 12606 Burton Street, North Hollywood, CA 91605. 213/768-3606.													•	
Hydrocap Corporation, 3579 East 10th Court, Hialeah, FL 33013. 305/696-2504.									•					•
Hyperion, Inc., 4860 Riverbend Road, Boulder, CO 80303. 303/449-9544.										•				
ILI, Inc., 5965 Peachtree Corners East, Atlanta, GA 30071. 404/449-5900.										•				
Illini Insulation & Sun Energy Conservation Experts, 906 South Cherry Street, P.O. Box 172, Effingham, IL 62401. 217/347-7935.										•				

International Business Services, Inc., 1010 Vermont Avenue Northwest, Washington, DC 20005. 202/789-5350.													•	
Intertechnology/Solar Corporation, 276 Broadview Avenue, Warrenton, VA 22186. 703/347-9500.	•										•		•	
Keystone Battery Corporation, 35 Holton Street, Winchester, MA 01890. 617/729-8333.									•					
Lafont Corporation, 1319 Town Street, Prentice, WI 54556. 715/428-2881.										•				
Largo Solar Systems, Inc., 991 South 40th Avenue, Ft. Lauderdale, FL 33317. 305/583-8090.												•		
Lennox Industries, Inc., P.O. Box 400450, Dallas, TX 75240. 214/783-5427.													•	
Lewis & Associates, 105 Rockwood Drive, Grass Valley, CA 95945. 916/272-2077.												•	•	
Midwest Components, Inc., Port City Industrial Park, Box 787, Muskegon, MI 49443. 616/777-2602.													•	
Multi-Fuel Energy Systems, 2185 North Sherman Drive, Indianapolis, IN 46218. 317/291-0400.											•			
National Semiconductors, Ltd., 331 Cornelia Street, Plattsburgh, NY 12901. 518/561-3160.	•													
Natural Power, Inc., New Boston, NH 03070. 603/487-5512.								•						
OEM Products, Inc., 2701 Adamo Drive, Tampa, FL 33605. 813/247-5947.										•				
Omnidata, Inc., 16 Springdale Road, Cherry Hill, NJ 08003. 609/424-4646.													•	
Omnium-G, 1815.5 North Orange-Thorpe Park, Anaheim, CA 92804. 714/879-8421.		•												
One Design, Inc., Mountain Falls Route, Winchester, VA 22601. 703/662-4898.											•		•	
Optical Radiation, 6352 North Irwindale Avenue, Azusa, CA 91702. 213/969-3344.													•	
Oriel Corporation of America, 15 Market Street, Stamford, CT 06902. 203/357-1600.													•	
Pinson Energy Corporation, P.O. Box 7, Marstons Mills, MA 02648. 617/428-8535.						•	•							
Propeller Engineering Duplicating, 403 Avenida Teresa, San Clemente, CA 92672. 714/498-3739.							•	•						
Ridgway Steel Fabricators, P.O. Box 382, Ridgway, PA 15853. 814/776-1323.										•				
Rothenberger, 5261 Edina Industrial Boulevard, Edina, MN 55435. 612/835-5550.												•		
Rotoflow Corporation, 2235 Carmelina Avenue, Los Angeles, CA 90064. 213/477-3083.		•		•										
Scotch Programs, P.O. Box 430734, Miami, FL 33143. 305/666-1300.													•	
Sennergetics, 18621 Parthenia Street, Northridge, CA 91324. 213/885-0323.													•	

Power generation/solar-related products

	Power generation systems and components								Other related systems and products					
	photovoltaics	solar thermal power generation	small hydraulic generators	ocean thermal energy conversion	space power satellite	wind generation systems	wind turbine generators	other wind generation components	electrical storage	heat recovery systems, components	biomass	solar tools and equipment	design aids, educational packages	factory-built homes
SES, Inc., Tralee Industrial Park, Newark, DE 19711. 302/731-0990.	•													
Shenandoah Manufacturing Company, Inc., P.O. Box 839, Harrisonburg, VA 22801. 703/434-3838.										•	•			
Silicon Sensors, Inc., Highway 18, East, Dodgeville, WI 53533. 608/935-2707.	•				•							•		
Siltec Corporation, 3717 Haven Avenue, Menlo Park, CA 94025. 415/365-8600.	•													
SJC Corporation, Division Frigiking-Tappan, 206 Woodford Avenue, Elyria, OH 44036. 216/329-2000.										•	•			
Small Hydroelectric Systems & Equipment, 15220 Sr 530, Arlington, WA 98223. 206/435-3148.			•					•						
SMC Energy Company, P.O. Box 246, Omaha, NE 68102. 402/397-1370.										•				
Solar Aquasystems, Inc., P.O. Box 88, Encintas, CA 92024. 714/753-0649.											•			
Solarcon, Inc., 607 Church, Ann Arbor, MI 48104. 313/769-6588.													•	
Solar Development, Inc., 3630 Reese Avenue, Riviera Beach, FL 33404. 305/842-8935.												•		
Solar-En Corporation, 3118 Route 10, Denbrook Village, Denville, NJ 07834. 201/361-2300.										•				
Solar Energy Corporation, 553 Pretty Brook Road, Princeton, NJ 08540. 609/924-1879.													•	

Company														
Solar Energy Products, Inc., Mountain Pass, Hopewell Junction, NY 12533. 914/226-8596.													•	
Solar Engines, 2937 West Indian School Road, Phoenix, AZ 85017. 602/274-3541.													•	
Solar Enterprises, Inc., 7928 N.E. Main Street, Fridley, MN 55432. 612/784-7177.										•				•
Solar Environmental Engineering, 2524 East Vine Drive, Ft. Collins, CO 80524. 303/221-5166.													•	
Solar Equipment Corporation, P.O. Box 357, Lakeside, CA 92040. 714/561-4531.		•												
Solar Equipment Distributors, Inc., Division of Yago Systems Design, P.O. Box 64, Barboursville, WV 25504. 304/736-4091.										•				
Solarex Corporation, 1335 Piccard Drive, Rockville, MD 20850. 301/948-0202.	•													
Solarflame Systems, Inc., P.O. Box 99, Leroy, IL 61752. 309/962-2861.										•				
Solar Home Systems, Inc., 8732 Camelot, Chesterland, OH 44026. 216/729-9350.													•	
Solar Industries, Monmouth Airport Industrial Park, Farmingdale, NJ 07727. 201/938-7000.													•	
Solar Kinetics, Inc., P.O. Box 47045, Dallas, TX 75247. 214/630-9328.	•													
Solaron Corporation, 720 South Colorado Boulevard, Denver, CO 80222. 303/759-0101.													•	
Solar Pathways, 3710 Highway 82, Glenwood Springs, CO 81601. 303/945-6503.													•	
Solar Power Corporation, 20 Cabot Road, Woburn, MA 01801. 617/935-4600.	•													
Solar Technology International, 119 North Center Street, Statesville, NC 28677. 704/873-7959.										•				
Solartherm, Inc., 1110 Fidler Lane, Silver Spring, MD 20910. 202/882-4000.	•										•			
Solartran Corporation, P.O. Box 496, Escanaba, MI 49829. 906/786-4550.														•
Solar Unlimited, Inc., 204 Oakwood Avenue, Huntsville, AL 35811. 205/534-0661.													•	
Solar World, Inc., 4449 North 12th Street, Suite 7, Phoenix, AZ 85014. 602/266-5686.													•	
Solatherm Corporation, 1255 Timber Lake Drive, Lynchburg, VA 24502. 804/237-3249.										•				
Solec International Inc., 12533 Chadron Avenue, Hawthorne, CA 90250. 213/970-0065.	•													
Solecon, P.O. Box 09763, 2770 East Main Street, Columbus, OH 43209. 614/236-5982.													•	
Solectro-Thermo, 1934 Lakeview Avenue, Dracut, MA 01826. 617/957-0028.	•													

Power generation/solar-related products

	Power generation systems and components								Other related systems and products					
	photovoltaics	solar thermal power generation	small hydraulic generators	ocean thermal energy conversion	space power satellite	wind generation systems	wind turbine generators	other wind generation components	electrical storage	heat recovery systems, components	biomass	solar tools and equipment	design aids, educational packages	factory-built homes
Sollos, Inc., 2231 Carmelina Avenue, Los Angeles, CA 90064. 213/820-5181.	•													
Solpub, Box 9209, College Station, TX 77840. 713/845-4133.													•	
Southeastern Solar Systems, Inc., 4705J Bakers Ferry Road. Atlanta, GA 30336. 404/691-1960.										•			•	
Spectrolab, Inc., 12500 Gladstone Avenue, Sylmar, CA 91342. 213/365-4611.	•												•	
S.P.S. Inc., 8801 Biscayne Boulevard, Miami, FL 33138. 305/754-7766.										•				
Structural Composites Industries, Inc., 6344 North Irwindale Avenue, Azusa, CA 91702. 213/334-8221.						•		•						
Sunshine Power Company, 1018 Lancer Drive, San Jose CA 95129. 408/446-2446.													•	
Sunspot Environmental Energy Systems, P.O. Box 5110, San Diego, CA 92105. 714/264-9100.												•		
Sun-Wind-Home-Concepts, Division of TWR Enterprises, 72 West Meadow Lane, Sandy, UT 84070.						•	•							
S. W. Energy Options, Route 8, Box 30-H, Silver City, NM 88061. 505/538-9598.											•			
T/Drill, Inc., 727 West Ellsworth, Building 8, Ann Arbor, MI 48104. 313/995-2187.												•		

Teledyne AERO-CAL, 528 East Mission Road, San Marcos, CA 92069. 714/744-1131.								•						
Teledyne Energy Systems, 110 West Timonium Road, Timonium, MD 21093. 301/252-8220.		•			•									
Thermo Control Wood Stoves, P.O. Box 640, Howe Caverns Road, Cobleskill, NY 12043. 518/296-8517.											•			
Tideland Signal Corporation, Environmental Energy Department, 4310 Directors Row, Houston, TX 77092. 713/681-6101.	•													
Topaz Electronics, 3855 Ruffin Road, San Diego, CA 92123. 714/279-0831.								•						
Trane Company, 3600 Pammel Creek Road, La Crosse, WI 54601. 608/787-3111.										•			•	
Unarco-Rohn, Division of Unarco Industries, Inc., P.O. Box 2000, Peoria, IL 61601. 309/697-4400.								•						
Vermont Iron Stove Works, Inc., The Bobbin Mill, Warren, VT 05674. 802/496-2617.											•			
Virginia Solar Components, Inc., Highway 29 South, Rustburg, VA 24588. 804/239-9523.										•				
West Wind Electronics, Inc., P.O. Box 542, Durango, CO 81301. 303/247-9677.						•	•	•						
Willey Corporation, 407 Palm Springs Boulevard, Indian Harbor Beach, FL 32937. 305/727-2040.													•	
WINCO, 7850 Metro Parkway, Minneapolis, MN 55420. 612/853-8400.						•								
Wind Engineering Corporation, P.O. Box 5936, Lubbock, TX 79417. 806/763-3182.						•								
Windworks, Inc., Box 44A, Route 3, Mukwonago, WI 53149. 414/363-4088.								•						
WM Lamb Company, P.O. Box 4185, 7116 Laurel Canyon, North Hollywood, CA 91607. 213/764-6363.	•													
Yankee Woodstoves, Box 7, Bible Hill Road, Bennington, NH 03442. 603/588-6358.											•			

Small Employers

There are too many small, local, individual companies to list them by name. This section covers the main types of small employers in enterprises that relate to energy-efficient shelter. Some of them may have only a few employees, and in some of the smallest, employees often take part in several aspects of the business—from manual labor to management.

To find the names of local employers, use the yellow pages of the telephone book, consult the Chamber of Commerce, or contact the nearest office of one of the organizations or government agencies listed in this part of the book.

Manufacturers

Many small firms manufacture materials needed to utilize and store solar energy. Workers may be employed in making the products, in managing the business, or in selling the products. Some of the products are:

collectors
hot water heaters
stoves
insulation
storm windows
swimming pool systems
air conditioning and refrigeration units

Retailers and Distributors

Building supply stores, lumber yards, fuel companies, and wholesalers sell the above products and employ people who can become knowledgeable about the properties, qualities, and relative advantages of the different types. Some of the jobs involve physically handling the material, purchasing from manufacturers, selling, and helping builders, or a combination of all four.

Utilities

Although large in capital investment, utilities are not large employers in local areas. They will be adding to their staffs, however, because now they are required by law to offer energy audits and to help consumers find ways to conserve energy.

Builders and Contractors

For those who enjoy physical labor, learning through building can be a good way to start out in the solar energy field. Every community has several builders, and plumbing, heating, and cooling contractors, but you need to read their advertisements carefully and talk to people for whom they have done work to identify the ones who are up-to-date on solar systems (active or passive) and energy conservation.

Architects, Designers, Consultants

Some architects and designers specialize in projects that utilize solar energy and good conservation measures. There are also consultants who have been in allied fields such as city planning who now help real-estate developers, government agencies, and other users of large-scale building complexes to incorporate energy-efficient features. When one of these professionals takes on a new project, he or she often will need extra staff, either people already trained in design, or those able to do research, contact suppliers, follow up on government regulations, write proposals, and so on. The American Institute of Architects can be helpful in identifying people with a special interest in solar.

Schools and Educational Groups

Many educational groups offer at least a brief series of seminars to inform the public about solar energy in homes, conservation measures, water and wind power, and other kinds of renewable energy. Such seminars take place in college continuing education divisions, public school adult education programs, science museums, and organizations concerned with the environment. Jobs become available when a school or organization needs someone to develop these offerings (often under special grants). Volunteering to work on educational projects can be a way of gaining knowledge and contacts.

Citizens' Action-Community Groups

It is hard to keep up with the almost daily growth of new grass-roots energy organizations. Some are loosely organized but enthusiastic groups that put out an occasional newsletter; others are essentially politically active and attempt to turn government policy toward renewable energy sources or small alternative technologies; still others are well-established organizations with regular publications and-or long-term projects. The jobs are often CETA—funded and usually the pay is low, but often the spirit and learning are high. Volunteers are welcome. People are needed to organize and run projects, to research and to write, and to produce visual material to reach the public with information about using renewable resources and conserving energy.

Education for Solar Technologies

University Programs and Research Groups

People who are interested in a professional career in research or in the design and use of solar energy systems should look into academic programs in architecture, mechanical engineering, electrical engineering, chemical engineering, physics, or chemistry. Basic principles in all these disciplines can be applied to solar energy. In addition, many colleges include aspects of solar energy in their regular technical courses, give a few special solar courses, and allow students to pursue their own interests in individual projects.

Prospective students may want to consider attending one of the colleges and universities that have well-developed solar energy programs. Here, students will find faculty and colleagues with solar energy interests, stimulating solar projects in process, and contacts that can help lead to the next step—graduate study or a job. The following list contains many academic institutions that have a strong interest in solar energy. They have laboratories and test facilities set up specifically for solar technologies, ongoing solar research by faculty members, and-or special solar curricula and degree programs.

United States

ALABAMA

University of Alabama, P.O. Box 1247, Huntsville, AL 35807.

School of Science and Engineering

Students may pursue graduate and undergraduate degrees in physics or mechanical engineering, and utilize all elective courses, both technical and nontechnical, to achieve concentrations in the various energy areas. General and engineering courses offered include ones on alternative energy sources.

Center for Environmental and Energy Studies

Research in many aspects of solar, wind, conservation, design and construction of buildings, storage, radiation, agricultural uses, and so

---

Practicing engineers, architects, and scientists who would like to become involved in solar energy research will find many of the University based research groups in this section. Research opportunities also exist in some of the private corporations (Part IV, "Private Corporations") and in the National Laboratories (Part IV, "Government").

forth; solar and wind equipment test facility for experimentation and certification; contracts to identify promising solar and wind sites across the country; study of renewable resources and self-sufficiency.

ALASKA

University of Alaska, Fairbanks, AK 99701.

Institute of Water Resources

Current interests of the professional staff include energy conservation, solar and alternative energy systems. The institute's laboratories and offices are available to graduate students, some of whom are on the research staff. Most of the professional staff hold appointments from more than one academic department and have a strong interest in graduate and undergraduate teaching.

ARIZONA

Arizona State University, Tempe, AZ 85281.

Solar energy research has been carried on at ASU since the early 1950s and some of its faculty members are internationally known in the field. A small group at the University from 1966 to 1970 later helped form the International Solar Energy Society. ASU has joint programs with government and industry, an interdisciplinary Committee on Solar Energy Research, and over 10,000 solar energy items in the library.

Department of Mechanical Engineering

Research in photovoltaics, parabolic reflectors, mirror solar concentrators. Research facilities include Direct Energy Conversion Laboratory, Mechanical Engineering Laboratory, solar simulators, computer modeling. Students may emphasize solar energy in their undergraduate technical electives or in graduate research.

College of Architecture

Research on heating and cooling; solar demonstration and experimental yard at the college; solar demonstration house. Undergraduate degree in solar studies, MA in solar technology offered.

Architectural Foundation

Did study on older solar homes in the United States for American Institute of Architects; library with solar energy index.

Laboratory of Climatology

Research in solar radiation, weather patterns.

Northern Arizona University, Flagstaff, AZ 86001.

College of Engineering and Technology

Research on solar heating and cooling, conservation in Department of Mechanical Engineering; elective courses in solar energy are available to students pursuing BS degrees in engineering and technology.

University of Arizona, Tucson, AZ 85721.

College of Engineering

Multi-departmental Energy Systems Engineering Program offers BS and MS degrees; solar energy engineering and optical science courses cover most aspects of theory and design: collectors, surfaces, cells, heating and cooling systems, storage, and so forth; laboratory facilities and independent study available.

Chemical engineering — Research on film for solar cells, solar energy for agricultural uses.

Electrical engineering — Research on photovoltaics, conservation in buildings.

Mechanical Engineering — Research on wind power, energy requirements of buildings.

Environmental Research Laboratory

Research in use of solar energy for food production, hot water and space heating; building design; bioconversion.

Optical Sciences Center

Graduate center for research in applied and theoretical optical physics includes solar energy systems such as heating and power generation, and economic analyses.

CALIFORNIA*

California Polytechnic State University, San Luis Obispo, CA 93407.

Department of Environmental Engineering

BS degree candidates can specialize in solar energy systems through several technical courses focusing on basic solar theory and applications.

---

* Many undergraduate colleges and universities in California offer one or two solar energy courses. A detailed listing of educational opportunities can be found in the *Educational Guide for Solar Energy and Energy Conservation at California Universities and Colleges*, Solar Energy Information Services, 18 Second Avenue, P.O. Box 204, San Mateo, CA 94401. $15.00

California State Polytechnic University, 3801 West Temple Avenue, Pomona, CA 91768.

Department of Mechanical Engineering

Courses in solar energy systems available to graduate and undergraduate students.

California State University at Long Beach, Long Beach, CA 90840.

Department of Mechanical Engineering

The department offers solar engineering courses and a BS degree in energy conservation.

California State University at Los Angeles, Los Angeles, CA 90032.

Department of Mechanical Engineering

The department offers several solar engineering courses for upperclass undergraduates.

California State University at Northridge, Northridge, CA 91330.

Departments of Mechanical and Chemical Engineering

The departments offer undergraduate courses covering general and engineering topics related to solar energy.

Northrup University, Inglewood, CA 90306.

Department of Energy Science

BS degree in science or in energy systems engineering includes many courses covering solar systems design, testing, evaluation, appropriate technology, alternative energy sources, photovoltaics, wind power, solar law, and legislation.

San Jose State University, San Jose, CA 95192.

Department of Environmental Studies

Students may pursue BA or BS degrees in environmental studies, with an emphasis in solar energy, through courses in solar theory, design, and engineering.

University of California at Berkeley, Berkeley, CA 94720.

Department of Engineering

The department offers a BS degree in solar engineering, with extensive solar-related courses in physics, electrical engineering, mechanical engineering, and computer science. Graduate level courses also available.

Lawrence Berkeley Laboratory, 1 Cyclotron Road, Berkeley, CA 94720.

Part of the National Laboratory system funded by the federal government, the Berkeley Laboratory is engaged in many aspects of research on energy-efficient buildings: computer analysis of energy use, analysis of passive solar heating and cooling designs, wall construction, lighting, and studies on the economic and sociopolitical impacts of regional energy development. The laboratory also conducts research on photosynthesis, cellulose conversion, and geothermal energy sources.

University of California at Davis, Davis, CA 95616.
Department of Atmospheric Science
Graduate Solar Energy Program is available to MA and MS degree candidates in atmospheric science. Courses cover radiation and other aspects of solar energy.

University of California at Los Angeles, Los Angeles, CA 90024.
Department of Architecture and Urban Design
Students in MA degree programs have available a number of courses in energy-conserving design, building climatology, and solar energy use and control.

University of California at San Diego, La Jolla, CA 92093.
Energy Center, La Jolla, CA 92037
Center develops courses in several departments at undergraduate and graduate levels dealing with all types of energy, including renewable sources. Research covers traditional sources, but some researchers are working on photovoltaics, energy conservation, structural designs, solar energy applications, and biomass production.

COLORADO

Colorado State University, Fort Collins, CO 80523.
College of Engineering
Faculty does not offer a degree in solar energy, believing that a strong basic engineering background gives the students the maximum options for the future, but the curriculum embodies the study of solar energy in several departments (civil engineering, mechanical engineering, electrical engineering, agricultural and chemical engineering, atmospheric science) at both the graduate and undergraduate levels.

Solar Energy Applications Laboratory

Conducts research in most solar technologies, including collectors, heating and cooling, conversion, wind, bioconversion, photovoltaics, and food processing. The professional staff runs workshops for contractors and engineers, and works with students doing research. The CSU Foothills Research Campus has four solar houses, one of which was the first solar-heated and cooled residential size building to be constructed. Faculty members are active in the field and some hold offices in the International Solar Energy Society. The National Science Foundation has funded many projects.

Colorado Technical College, Colorado Springs, CO 80907.

Solar Engineering Technology Department

Offers BS degree, with several courses in theory, design, and evaluation of solar systems.

University of Colorado at Boulder, Boulder, CO 80309.

Department of Environmental Design

BA and MA candidates in environmental design can specialize in solar energy and appropriate technology. Solar-related courses are offered in departments of mechanical, chemical, and electrical engineering, physics, astrogeophysics, and arts and sciences. Laboratories for astrophysics and for atmospheric and space physics are part of the program. Faculty is engaged in research in most solar technologies, including building construction, space and water heating, radiation, and storage.

University of Colorado at Colorado Springs, Colorado Springs, CO 80907.

Department of Physics and Energy Sciences

Offers Distributed Studies in Energy Sciences Program for BS degree, with courses covering solar energy and conservation, appropriate technology, solar systems design and evaluation.

DELAWARE

University of Delaware, Newark, DE 19711.

Institute of Energy Conversion, 1 Pike Creek Center, Wilmington, DE 19808.

Research in energy conservation, photovoltaics, storage systems, collectors, space and water heating, radiation; built Solar House I (experimental solar laboratory demonstrating solar applications for domestic use) on the Newark campus. Some graduate and undergraduate students participate in the research programs.

DISTRICT OF COLUMBIA

George Washington University, Washington, DC 20052.
Department of Civil, Mechanical, and Environmental Engineering
Research on energy storage, collectors, solar heating and cooling. Department has graduate program in Energy Resources and Environment, and courses in solar and wind energy.

FLORIDA

Florida Institute of Technology, P.O. Box 1150, Melbourne, FL 32901.
Department of Mechanical Engineering
Several courses on solar energy conversion systems are available to graduate students; the Center for Research is also involved in solar research.

Florida International University, Miami, FL 33199.
Department of Technology
The department offers the MS in solar energy technology and has several courses in solar design, testing, and evaluation.

Florida Solar Energy Center, 300 State Road 401, Cape Canaveral, FL 32920.
The center is part of the state university system of Florida. It does research and testing and develops standards in several solar technologies, provides technical assistance to manufacturers, and funds solar research and development in the state public and private universities.

University of Florida, Gainesville, FL 32611.
Department of Mechanical Engineering
Candidates for MS degrees have opportunity to do research and take courses in wind, methane, greenhouses, and other solar technologies.

University of Miama, Coral Gables, FL 33124.
School of Engineering and Architecture
Graduate and undergraduate students in mechanical engineering may elect energy engineering option, which includes courses and research in solar energy. The Clean Energy Research Institute in this school is the focal point of energy-related activities at the university.

The professional staff is composed of faculty from all scientific and engineering disciplines and conducts research on many aspects of solar energy. The institute also runs regional, national, and international conferences, some of a highly technical nature, for scientists and researchers.

GEORGIA

Georgia Institute of Technology, Atlanta, GA 30332.

Georgia Tech has a large solar energy research and development program involving faculty from all disciplines and also extensive solar test facilities on campus. Work is carried on in the use of solar energy for heating and cooling large buildings, thermal power plants, industrial processes, and agricultural drying processes.

College of Architecture

Faculty are involved in research on solar design, construction of buildings, and energy conservation. The college stresses education and training at all levels and ages, including continuing education.

College of Engineering

Also involved in research, the department offers a multi-disciplinary energy engineering specialization that includes solar energy at both the graduate and undergraduate levels.

The Technology and Development Laboratories at the Engineering Experiment Station have advanced technological facilities for research in biocombustion and passive solar. Graduate students may have research assistanceships to participate in research and pursue a thesis topic.

HAWAII

Hawaii Natural Energy Institute, 2540 Dole Street, Holmes 240, Honolulu, HI 96822.

Part of the University of Hawaii, the institute is well supported by the county, the state, and the U.S. Department of Energy. It does extensive research on five natural energy resources: geothermal, ocean, biomass, wind, and direct solar energy, and is preparing a program aiming for Hawaii County energy self-sufficiency by 1990. The research involves both graduate and undergraduate students in the scientific and engineering disciplines at the University of Hawaii. MS and PhD candidates produce theses and dissertations on topics connected with the institute activities.

IDAHO

University of Idaho, Moscow, ID 83843.

Courses covering solar topics are offered in several departments: architecture, electrical engineering, mechanical engineering, and industrial education.

ILLINOIS

Illinois State University, Normal, IL 61761.

Department of Industrial Technology

Undergraduates may take energy option, with courses in solar heating and cooling and conservation.

University of Chicago, Chicago, IL 60637.

Department of Physics

Faculty is engaged in research and there are research opportunities for graduate level students in photovoltaics and collector design and evaluation.

University of Illinois, Chicago Circle, Chicago, IL 60680.

Department of Architecture and *Energy Resources Center*

Conduct research in energy conservation, solar building design, space and water heating, and energy efficiency in small industries. Some solar courses are given. Energy Engineering Program is offered by Engineering Department.

At the Urbana campus, solar topics are covered in courses in chemical and mechanical engineering, architecture, and geology.

IOWA

University of Iowa, Iowa City, IA 52242.

Division of Energy Engineering

Research in thermal storage and collectors is carried on in the division's laboratory test facilities. Graduate students specializing in thermal sciences may participate in solar research.

Department of Mechanical Engineering

Undergraduates may choose solar energy electives, and candidates for graduate degrees may concentrate in area of solar energy.

KANSAS

Kansas State University, Manhattan, KS 66502.

Department of Agricultural Engineering

Faculty engaged in research on many aspects of energy use in agriculture, including conservation of energy in agricultural methods, solar heating for animal shelter, and grain drying.

Department of Architecture

Several undergraduate courses cover conservation and use of solar energy systems.

Department of Mechanical Engineering

Department offers senior level electives in solar thermal processes and has additional courses and a solar energy research laboratory under development.

Center for Energy Studies in College of Engineering does research in solar hot water systems, collectors, crop residue as energy source, use of alternative energy in low-income homes, development of Kansas Conservation Plan, citizens' workshops on energy, secondary school lectures, and material on energy.

KENTUCKY

Western Kentucky University, Bowling Green, KY 42101.

Department of Engineering Technology

Study for BS in engineering technology can include courses in solar fundamentals, building technology, and energy conservation.

LOUISIANA

Louisiana State University, Baton Rouge, LA 70803.

Solar Energy Program

The university has been delegated the responsibility for the state solar energy program, Solar Energy Awareness in Louisiana (SEAL). The program involves research, demonstration, and promotion of solar energy. Solar heating and hot water systems have been installed on the LSU field house. The Department of Mechanical Engineering does research in solar heating systems.

MARYLAND

University of Maryland, College Park, MD 20740.

Department of Mechanical Engineering

Department has Solar Energy Laboratory, does research in cooling, heating, hot water systems, and industrial process heat. Faculty members also run conferences and workshops. Graduates and undergraduates take courses in engineering of solar energy, participate in experiments. Graduate assistanceships available.

MASSACHUSETTS

Boston University, Boston, MA 02215.

Department of Chemistry

Research in photochemistry and photovoltaics.

Center for Energy Studies, 195 Bay State Road, Boston, MA 02215. Does research and provides education for the public on renewable resources and conservation of energy.

University of Lowell, Lowell, MA 01854.

The University offers MS and PhD degrees in physics, with a solar energy option, and BS and MS degrees in engineering, with an emphasis on solar. Several solar courses are available.

University of Massachusetts, Amherst, MA 01003.

Department of Mechanical Engineering

Research in wind energy, including facilities with model wind furnace, wind and solar combinations for space heating. Department offers a BS with an energy option.

Massachusetts Institute of Technology, Cambridge, MA 02139.

Department of Chemistry

Research in photochemistry.

Departments of Chemical Engineering and Mechanical Engineering

Research in heating and cooling; helped design Solar House IV (experimental and demonstration building).

School of Architecture and Planning

Research in heating, building design; built Solar House IV.

Research programs are well-funded by government and industry, and many faculty members are involved individually.

Graduate students may participate in solar research in the various departments. Solar energy is covered as part of regular undergraduate courses in several departments: electrical, chemical, and mechanical engineering, architecture, materials science, computer science. Students at MIT recently formed their own Appropriate Technology Group.

Northeastern University, Boston, MA 02115.
Departments of Mechanical Engineering and Chemical Engineering offer a number of solar courses for graduate and undergraduate engineering students.

Worcester Polytechnic Institute, Worcester, MA 01609.
Education at WPI has a unique structure based on individual projects carried out by students, with faculty supervision. From 60 to 80 solar energy projects, involving one to three students each, are carried on each year. Faculty from several departments do solar-related research, especially in photovoltaics and solar hot water systems.

MICHIGAN

University of Detroit, Detroit, MI 48221.
Department of Mechanical Engineering
BS degree candidates have opportunity for cooperative program with Jordan College Energy Institute (see below).

Jordan College, 360 West Pine Street, Cedar Springs, MI 49319.
Energy Division
The division was created as a result of response to Jordan College's early activity in solar installations and alternative energy academic programs. Students may complete an Associate Degree in alternative energy at Jordan and then transfer to obtain four-year engineering degrees: BS in mechanical engineering at the University of Detroit or BS in engineering technology at Wayne State University or Saginaw Valley College.

Jordan Energy Institute
Provides a one-semester, full-time concentration of five alternative energy courses for 15 hours of college credit. Core courses cover solar, wind, geothermal, hydroelectric, and biomass, with laboratory, classroom, and field experience. The institute program is part of a sequence of alternative energy courses for Jordan College students in Associate Degree or BA programs, but it is also available to students from other colleges. (See also "Special College Programs.")

University of Michigan, Ann Arbor, MI 48104

Department of Mechanical Engineering

Does research in heating, cooling, and power generation. BS or MS candidates can concentrate in the energy conversion and power field, which includes work in solar heating, cooling, and power generation. Courses also are offered in solar fundamentals and systems design.

Saginaw Valley College, University Center, Saginaw, MI 49701.

Department of Mechanical Engineering Technology

BS degree candidates have opportunity for cooperative program with Jordan College Energy Institute (see page 171).

Wayne State University, Detroit, MI 48202.

College of Engineering Energy Center

Does research in alternative fuels, conservation techniques, alternative energy sources. Advanced undergraduates and graduate students may participate.

Department of Mechanical Engineering Technology

BS degree candidates have opportunity for cooperative program with Jordan College Energy Institute (see page 171).

MINNESOTA

University of Minnesota, Minneapolis, MN 55455.

Department of Mechanical Engineering

Research in heat transfer, storage, coatings. BS, MS, and PhD candidates may specialize in solar energy through electives and research projects.

School of Architecture and Landscape Design

Research in heating, cooling, collectors, and earth-sheltered design. Project Ouroboros studies the architect's role in energy conservation through research and enables students to design and construct two full-scale residences that demonstrate total energy conservation. Solar research projects at the university often are funded by government grants, and by some work in cooperation with Honeywell Systems and Research Center.

MISSISSIPPI

Mississippi State University, Mississippi State, MS 39762.

Department of Mechanical Engineering

Research in conservation, wind, water and space heating, storage.

Courses in solar energy thermal processes are available in graduate and undergraduate programs.

Energy Research Center

Part of the College of Engineering, the center does solar collector design and energy modeling for the state.

MISSOURI

University of Missouri at Columbia, Columbia, MO 65201.

Department of Energy Systems and Resources

Department offers BS degree. Several courses in solar engineering are given by this department and the Department of Mechanical and Aerospace Engineering.

University of Missouri at Rolla, Rolla, MO 65401.

Department of Mechanical Engineering

Research in heating, energy conversion. Solar Energy Conversion Program available to graduate students.

Department of Electrical Engineering

MS offered with Solar Energy Conversion Program.

Southwestern Missouri State University, Springfield, MO 65802.

Department of Physics

BS degree candidates may take engineering physics—Solar Emphasis, including several technical and laboratory courses in solar systems.

Washington University, St. Louis, MO 63130.

Department of Mechanical Engineering

Has large solar test facility for research in collectors, solar systems analysis and monitoring, design of active and passive systems, computer simulation of systems, and economic analyses of solar systems. Department has materials research and other laboratories, and an interdisciplinary Center for Development Technology that performs planning and policy studies, including renewable resource utilization. Graduate and undergraduate students participate in research as technicians and research assistants.

Department of Technology and Human Affairs

Graduate and undergraduate degrees in engineering and public policy provide professional engineering education, coupled with knowledge of contemporary problems and issues that involve technology (see "Special College Programs").

NEVADA

University of Nevada, Boulder City, NV 89005.

Desert Research Institute

Energy Systems Center does research and development in solar, wind, energy storage. Faculty have joint appointments at institute and university campuses and supervise students at the graduate level in various scientific and engineering fields.

University of Nevada, Las Vegas, NV 89154.

Department of Engineering

Research projects in solar, wind, geothermal energy, and formation of building codes that include solar construction. Students may elect to concentrate in technical areas involving solar energy.

NEW HAMPSHIRE

Dartmouth College, Hanover, NH 03775.

Thayer School of Engineering

Research in heating, photovoltaics. Graduate and undergraduate degree programs allow students to develop special engineering design problems that may include solar energy. Courses are available in solar engineering design.

NEW JERSEY

Glassboro State College, Glassboro, NJ 08028.

Industrial Education and Technology Department

Department has Energy and Transportation Program, with courses in solar energy. Awards BA and MA degrees.

Montclair State College, Upper Montclair, NJ 07043.

Industrial Education and Technology Department

BA and BS degree programs have Industrial Powers Curriculum, with courses in wind and other alternative energy conversion systems.

New Jersey Institute of Technology, Newark, NJ 07101.

Department of Mechanical Engineering

Has Mechanical Engineering and Technology Program, with courses in solar heating design and other solar applications, and awards BS degree.

Princeton University, D329 Engineering Quadrangle, Princeton, NJ 08540.

Department of Mechanical Engineering

PhD, MS, BS, and BA degrees offered with an Energy Conversion and Resources Program. A number of courses include solar engineering material.

Center for Environmental Studies

Research in solar, wind, methane, conservation; has Ford Foundation grant to study feasibility of decentralized organization of solar energy; acts as focal point for students engaged in multidisciplinary research.

School of Engineering and Applied Science

Research has included energy conservation experiments and development of low-cost wind power systems.

School of Architecture and Urban Planning

Research in solar heating.

Ramapo College of New Jersey, Mahwah, NJ 07430.

Department of Environmental Studies

Offers BS and BA degrees and Alternative Energy Curriculum, with several courses.

Rutgers University, Piscataway, NJ 08854.

Department of Electrical Engineering

Research on solar energy storage.

Department of Biological and Agricultural Engineering

Cook College, P.O. Box 231, New Brunswick, NJ 08903

Research in solar heating and in reducing heat requirements of greenhouses by using low-cost solar collectors and movable insulating blankets.

Stockton State College, Pomona, NJ 08240.

Natural Sciences Division

Research in solar, wind; development of data recording system and cost comparisons of various heating systems; and analyses of collector design.

NEW MEXICO

New Mexico State University, Las Cruces, NM 88003.

The university has built several solar buildings and scientists have been active in solar research here since the 1940s. Some faculty are involved in solar activity at the national and international level.

New Mexico Solar Energy Institute

Full-time staff of scientists and engineers are doing extensive research on wind, geothermal, biomass, solar thermal power, heating, cooling, hot water systems, and building design. Institute also tests heating and cooling systems, coordinates alternative energy research within the state, and acts as a resource for information.

Department of Mechanical Engineering

Research in energy conversion, solar, wind. MS or BS degree students can specialize in solar engineering. Windmill, solar-heated and -cooled house are among installations available for study and research; technical solar courses also offered.

University of New Mexico, Albuquerque, NM 81731.

Department of Architecture and Planning

Graduate and undergraduate students may do internships or participate in solar heating and cooling, energy conservation projects at the Center for Environmental Research and Development. The department has a stated emphasis on solar design and appropriate technology.

Department of Mechanical Engineering

Research in heating and cooling, wind energy.

Department of Chemical Engineering

Research in geothermal energy. Graduate assistanceships available in these departments; both faculty and students are involved in research.

Technology Application Center

Publishes surveys and bibliographies and gives short courses on alternative energy topics.

NEW YORK

Adelphi University, Garden City, NY 11530.

Department of Physics

Offers MS and BS degrees in energy studies, with technical solar courses available.

City University of New York (CUNY), New York, NY 10031.

The Physics and Mechanical Engineering Departments offer some undergraduate solar courses at the Brooklyn, Staten Island, and New York City campuses. CUNY Graduate School and University Center in New York City offers graduate level solar courses in mechanical and electrical engineering.

Manhattan College, Bronx, NY 10471.

Department of Engineering

Has several courses at the graduate level, covering solar energy conversion systems and radiation.

Rensselaer Polytechnic Institute, Troy, NY 12181.

BS degree candidates in architecture and engineering have available technical courses that cover solar technologies.

Rochester Institute of Technology, 1 Lob Memorial Parkway, Rochester, NY 14623.

Department of Mechanical Engineering

BS and MS degree work can include solar energy research carried on at Energy House, a solar house located on the campus.

State University of New York (SUNY), Albany, NY 12222.

The SUNY system has four university centers and a total of 64 campuses, many of which give a few solar courses. Some locations, however, have placed an emphasis on education for solar technologies.

SUNY at Albany, Albany, NY 12222.

Department of Atmospheric Sciences

Provides courses in solar radiation at the graduate level. The Atmospheric Sciences Research Center is the training site for monitoring solar radiation and meteorology. A solar-heated "minimum energy building" was constructed in collaboration with students and volunteers from the community.

SUNY at Binghamton, Binghamton, NY 13901.

Department of Physics

BA and BS degree candidates may take specialization in solar energy, with several courses covering solar technologies.

SUNY at Buffalo, Buffalo, NY 14260.

Department of Engineering

Many technical solar courses available to graduate and undergraduate students in engineering science, electrical, chemical, and mechanical engineering. Energy conversion and photovoltaics are stressed.

SUNY at Plattsburgh, Plattsburgh, NY 12901.

Department of Environmental Science

Offers BA in environmental science, alternative energy, with solar-related courses in several departments.

Syracuse University, Syracuse, NY 13210.

A Selected Studies Plan allows individually designed majors at the graduate and undergraduate levels in the College of Arts and Sciences. There are also All-University Centers for Multi-Disciplinary Research. Using this structure, students may specialize in solar technologies.

School of Architecture

Gives students working toward graduate and undergraduate degrees an opportunity to concentrate in alternative energy use and energy-conserving design.

NORTH CAROLINA

Duke University, Durham, NC 27706.

Department of Engineering

Offers BS and MS degrees with Energy Conservation Program; courses cover most solar technologies.

University of North Carolina.

Campuses at Asheville, Charlotte, and Greensboro offer technical solar courses through Departments of Physics and Engineering.

NORTH DAKOTA

University of North Dakota, Grand Forks, ND 58202.

Engineering Experiment Station

Under contract to Oak Ridge National Laboratory, and with additional support from a number of industries, the Experiment Station designed, constructed, and monitors a solar house in Larimore, North Dakota. Started in 1976, the research continues to improve the components and the total system for active solar heating.

OHIO

University of Cincinnati, Cincinnati, OH 54221.

Undergraduate students in architecture or mechanical engineering technology have available a number of courses covering alternative energy systems, energy-conscious design of buildings, solar heating and cooling.

University of Dayton, 300 College Park Avenue, Dayton, OH 45469.

Department of Mechanical Engineering

Does research on liquid collectors and testing of standard solar collectors. Undergraduate students participate by testing collectors on roof of Engineering Laboratories Building. Courses in energy conversion for MS and BS degree candidates cover solar technologies.

Research Institute

University's solar energy research programs, concentrating on heating and cooling of buildings and on wind energy conversion, are carried out jointly by institute and Mechanical Engineering Department. Solar heating systems were designed for four large nonresidential buildings. Institute work provides laboratory opportunities for undergraduate engineering students, research projects for graduate students.

Youngstown State University, Youngstown, OH 44555.

Department of Electrical Engineering

Several solar energy engineering courses are offered.

OKLAHOMA

Oklahoma State University, Stillwater, OK 74074.

School of Technology

Research in earth coils and geothermal wells used as solar energy storage devices.

Department of Electrical Engineering

Graduate and undergraduate students may take energy option; courses in energy conversion cover solar technologies.

Department of Architecture

Graduate students have short courses and regular courses available covering energy conservation, earth-sheltered design, alternative energy systems.

University of Oklahoma, Norman, OK 73019.

School of Aerospace, Mechanical, and Nuclear Engineering

Research on wind energy systems, solar collectors, and solar stock tank warmer for agricultural use.

Department of Architecture

Undergraduate courses stress energy conservation and passive solar design.

University of Tulsa, Tulsa, OK 74104.

Division of Resource Engineering

Faculty engaged in solar energy projects involving heating systems, collectors, and storage. Solar energy engineering courses available at undergraduate level.

OREGON

Oregon State University, Corvallis, OR 97331.

Department of Mechanical Engineering

Faculty involved in research at Solar Energy Center, University of Oregon.

Department of Atmospheric Sciences

In BS, MS, and PhD programs, courses available in solar radiation and meteorological measurements, with laboratory experience at the Solar Radiation Measurement Facility.

Office of Energy Research and Development

Office is working with faculty to fund and develop research and training programs related to conservation of energy and new energy sources.

University of Oregon, Eugene, OR 97403.

Solar Energy Center

Research on heating, cooling, and hot water systems, passive solar design, development of collector-reflector combinations, study of legal, economic, and other technical problems that accompany solar energy development. Center emphasizes regional approach to solar development; issues publications and runs seminars.

Department of Architecture
Department of Physics

Faculty members do research in the Solar Energy Center and teach solar technologies in courses offered to undergraduates in their departments. Students are encouraged to pursue research interests at the center.

PENNSYLVANIA

Pennsylvania State University, University Park, PA 16802.

Architectural Engineering Department

Research, with instrumentation of residence to measure heat loss with infiltration; building and modeling of energy systems to reduce heat loss; design and construction of greenhouses. Offers undergraduate course in solar energy building system design.

University of Pennsylvania, Philadelphia, PA 19174.

The Energy Center

Engineering faculty members are involved in research in several solar technologies and graduate courses covering solar energy topics are offered in several departments: Systems Engineering, Mechanical Engineering, Electrical Engineering, and Materials Science Departments.

PUERTO RICO

University of Puerto Rico.

Center for Energy and Environment Research, Caparra Heights Station, San Juan, PR 00935

Center is focal point of alternative energy research in Puerto Rico and is an operating unit of the university, working with all the campuses and subdivisions of the system. Extensive research is conducted in solar thermal, photovoltaics, bioconversion, wind, ocean thermal, and cogeneration.

RHODE ISLAND

Brown University, Providence, RI 02912.

Department of Engineering

Faculty does research in photovoltaics, collectors; offers graduate course in photovoltaics.

University of Rhode Island, Kingston, RI 02881.

Department of Mechanical Engineering

Department does research in solar collectors, has graduate degree programs specializing in thermal science, including solar and wind energy. Courses also available in energy conservation and building design.

Center for Energy Study

Established in 1977, center conducts energy-related research, offers technical advice, and runs educational programs on energy conservation.

SOUTH CAROLINA

Clemson University, Clemson, SC 29631.

Graduate and undergraduate engineering students may specialize in energy systems; alternative energy, energy conversion, conservation, and other solar technologies covered in the courses of several departments.

University of South Carolina, Columbia, SC 29208.

College of Engineering

Does research in water and space heating; offers an undergraduate solar course in Department of Mechanical Engineering.

SOUTH DAKOTA

South Dakota State University, Brookings, SD 57007.

Department of Mechanical Engineering

Research in solar heating for farm buildings and for grain drying. Graduate and undergraduate students may take technical electives in solar energy; sequence of several mechanical engineering courses in heating, cooling, refrigeration, and design of thermal systems was updated in 1979 to concentrate on energy efficiency in buildings and use of solar energy. Many students choose solar-related projects in laboratory courses.

Department of Agricultural Engineering

Research on farm heating needs.

Department of Electrical Engineering

Research in wind power.

TENNESSEE

Tennessee Technological University, Cookeville, TN 38501.

Department of Mechanical Engineering

Research in collectors and heating systems. Undergraduate students may take course in solar energy processes and systems, and choose solar specialized option in senior year. Graduate degree programs also offered.

University of Tennessee, Knoxville, TN 37916.

Department of Mechanical and Aerospace Engineering

Faculty does research in heating, storage, energy conversion, and offers undergraduate course in solar energy utilization.

Environment Center

The center was created to support interdisciplinary study of alternative solutions to problems related to energy and the environment. Graduate students and faculty members participate in research and development projects on solar energy utilization and energy conservation in buildings and in industry.

TEXAS

American Technological University, Killeen, TX 76541.

The American Section of the International Solar Energy Society has its headquarters here and many of the faculty are involved in solar research and development.

Department of Management and Business

Department offers MS in energy management sciences, which covers scientific and engineering, as well as management topics. Specialized courses deal with legislative, economic, and management aspects of solar energy, in addition to courses on wind, passive design, heating and cooling systems, storage, energy conversion, computer simulation, and agricultural applications.

Rice University, Houston, TX 77001.

The Graduate School of Engineering has a strong research orientation in engineering fundamentals and offers a Space Solar Power Research Program. The Departments of Mechanical Engineering and Materials Science and of Space Physics give some undergraduate solar courses.

Southern Methodist University, Dallas, TX 75275.

Departments of Engineering

Regular engineering courses cover solar-electric conversion and use of solar energy in heating and cooling systems. Some special courses are available at graduate and undergraduate levels in photovoltaics and solar energy applications.

Texas A & M University, Main Campus, College Station, TX 77843.

Department of Mechanical Engineering

Research in all main solar technologies and computer facilities for simulating systems. Graduate and undergraduate degree candidates also have available courses dealing specifically with solar energy applications and processes.

Department of Architecture

Research in energy-conscious building design; graduate level courses in conservation, alternative architecture, and use of solar systems.

Trinity University, 715 Stadium Drive, San Antonio, TX 78284.

Faculty of Sciences, Mathematics, and Engineering

Offers MS in solar energy studies or applied solar energy to graduate students majoring in engineering, physics, or computer science; individualized programs also available. The Solar Energy Graduate Program is administered by an interdisciplinary faculty committee and includes courses in materials and design of solar systems, collectors, radiation, and computer modeling. Research assistants participate in departmental research. The university has the nation's largest Federal Solar Heating and Cooling Commercial Demonstration Project, used by students in the Solar Energy Graduate Program. It also operates the Department of Energy six-state Solar-Meteorological Research and Training Program.

University of Houston, Houston, TX 77004.
Energy Laboratory
The laboratory is the focus of research, development, and evaluation of solar energy systems. Research is conducted in biomass, materials for receiving and converting solar energy, collectors, conservation in buildings and in industrial processes, chemical systems for storage, and solar thermal power systems. Studies also are done on public policy and economic aspects of energy. Faculty from the engineering, science, and economics disciplines are affiliated with the laboratory.

University of Texas at Austin, Austin, TX 78712.
Center for Energy Studies
Research on geothermal, solar energy conversion. A number of courses offered at the undergraduate level in the Departments of Architecture and Mechanical Engineering cover energy-conscious design, heating and cooling, and solar thermal systems.

University of Texas at Dallas, Richardson, TX 75080.
Department of Environment Sciences
Department has Solar Energy in Environment Studies for MS and PhD degree candidates. A number of graduate level courses are offered in solar technologies, energy conversion, and alternative energy resources.

UTAH

University of Utah, Salt Lake City, UT 84112.
Department of Mechanical Engineering
Research on space heating, geothermal; energy research laboratories include an active solar energy facility. Several undergraduate courses cover solar building design and thermal applications of solar energy.

Utah State University, Logan, UT 84322.

College of Engineering

Research in bioconversion, collectors, storage, and agricultural applications.

Department of Mechanical Engineering

Special courses and individual experimental projects enable graduate and undergraduate students to specialize in energy systems, including solar, wind, bioconversion, and geothermal.

Department of Biometeorology

Solar technology is covered in several courses at the graduate level.

VIRGINIA

Old Dominion University, Norfolk, VA 23508.

Department of Mechanical Engineering

A number of undergraduate courses are offered in solar technologies and students may take power-energy conversion option or Solar Energy Systems Program. Graduate degrees also awarded.

University of Virginia, Charlottesville, VA 22901.

Department of Mechanical Engineering

Research in heating and cooling, collectors. Courses and group projects in solar system design are available.

Virginia Polytechnic Institute and State University, Blacksburg, VA 24061.

College of Architecture and Urban Studies

Courses in building technology cover energy conservation and solar heating and cooling systems.

Agricultural Engineering Department

Research in wind, building design, and energy conservation.

WASHINGTON

Joint Center for Graduate Study, 100 Sprout Road, Richland, WA 99352.

Research in photovoltaics. Research assistanceships are available to graduate students from Oregon State University, Washington State University, and the University of Washington.

Washington State University, Pullman, WA 99163.

Department of Mechanical Engineering

Research in solar water heating, solar energy, and peak load considerations of utilities, space heating, collectors. Solar Energy Laboratory recently constructed and its use by students is encouraged.

Department of Architecture

Students participate in research on test buildings having solar passive design.

University of Washington, Seattle, WA 98195.

Institute for Environmental Studies

Faculty from Departments of Atmospheric Sciences, Mechanical Engineering, Civil Engineering, Architecture, and Forest Resources participate in research and development project "the Incorporation of Energy Conservation Principles Into the Design of State Buildings." Student research assistants are involved.

Department of Architecture

Research in lighting, earth-covered structures. Undergraduate courses stress energy-conscious design and solar energy applications in buildings.

WEST VIRGINIA

West Virginia University, Morgantown, WV 26505.

Department of Aerospace Engineering

Research in wind energy systems. Courses are available in this department and in the Department of Technology Education in appropriate technology and solar energy systems.

WISCONSIN

Marquette University, 1515 West Wisconsin Avenue, Milwaukee, WI 53233.

Department of Electrical Engineering

Concentration in energy system engineering includes solar and conservation topics, with specific courses in solar energy engineering for graduate and undergraduate students.

Department of Mechanical Engineering

Concentration in energy engineering may include solar courses and independent study at graduate and undergraduate levels.

University of Wisconsin at Madison, Madison, WI 53706.

Solar Energy Laboratory

Started in 1953, with emphasis on solar energy for the developing countries, the laboratory is now a leader in research, particularly in solar energy computer simulation. Faculty members of the Chemical and Mechanical Engineering Departments staff the laboratory and supervise MS and PhD degree candidates doing theses in aspects of solar energy utilization. The director of the laboratory is a past president of the International Solar Energy Society. Several other specialized laboratories are available to students in the engineering departments.

Department of Mechanical Engineering

Graduate and undergraduate courses cover appropriate technology, conservation, solar energy conversion, and other solar applications.

University of Wisconsin at Milwaukee, Milwaukee, WI 53201.

Department of Architecture and Urban Planning

Department offers a Solar Architecture Program at the undergraduate level, with a number of courses that cover energy-conscious design, design of solar buildings, and use of solar systems.

WYOMING

University of Wyoming, Laramie, WY 82071.

Office of Research

Research on solar, wind, and socioeconomic impact of energy development on Wyoming. Courses in Departments of Engineering, Natural Science, and Geography include material on solar technologies.

Canada

BRITISH COLUMBIA

University of British Columbia, Vancouver, BC Canada V6T 1W5.

Department of Mechanical Engineering

Offers graduate and undergraduate courses in solar energy utilization.

NEW BRUNSWICK

University of New Brunswick, Box 4400, Fredericton, NB, Canada E3B 5P7.

Department of Mechanical Engineering

Graduate and undergraduate courses in solar energy cover water and space heating, photovoltaics.

ONTARIO

Lakehead University, 855 Oliver Road, Thunder Bay, Ontario, Canada P7B 5E1.

Department of Physics

Energy and fuel science courses cover solar technologies.

Carleton University, 1231 Colonel by Way, Ottawa, Ontario, Canada.

Faculty of Engineering

Undergraduate and graduate courses in energy conversion and power generation include solar energy.

Laurentian University, Ramsey Lake Road, Sudbury, Ontario, Canada P3E 2C6.

Department of Physics

Solar energy is an important part of several regular courses in the curriculum.

McMaster University, Hamilton, Ontario, Canada L8S 4M1.

Engineering Physics Department

Research on solar cells, storage systems. Courses in physics and in optoelectronics cover solar technologies.

University of Waterloo, Waterloo, Ontario, Canada N2L 3G1.

Department of Mechanical Engineering

Special solar energy courses at graduate and undergraduate level.

University of Western Ontario, 1151 Richmond, London, Ontario, Canada N6A 5B9.

Faculty of Engineering

Advanced undergraduate courses offered in solar energy utilization and solar applications in energy conversion.

QUEBEC

Concordia University, 14-15 DeMaisonneuve Boulevard West, Montreal, Quebec, Canada H3G 1M8.

Graduate and undergraduate courses in solar equipment design and principles of solar engineering.

Macdonald College, Ste. Anne de Bellevue, Quebec, Canada H0A 1C0.

Brace Research Institute

Research in most solar technologies including wind, water and space heating, radiation, agricultural uses, storage, small equipment (appropriate technology).

SASKATCHEWAN

University of Saskatchewan, Saskatoon, Saskatchewan, Canada S7N 0W0.

Department of Mechanical Engineering

Research in collectors, design and construction of buildings suitable for cold climates, heat exchangers, energy conserving greenhouses. Regular courses cover solar technologies and the department has a special course in solar engineering.

Department of Chemical Engineering

Research in bioconversion.

Department of Electrical Engineering

Research in photovoltaics.

Department of Agricultural Engineering

Research in heat recovery systems for livestock housing.

Two-Year Degree Programs for Solar Technicians

Solar technicians are the people who install, operate, and maintain the systems that use solar energy. Although they may work with research teams or in demonstration projects, most of them work at the local level on solar equipment for heating hot water and for heating and cooling residential and commercial buildings. As they gain experience, such technicians often choose the components and design the systems, as well as installing and repairing them.

A number of private and community colleges now offer programs in solar system design, installation, and maintenance. The training involves many of the skilled trades—carpentry, plumbing, electrical wiring, sheet metal work—in addition to courses in solar principles.* Some of the basic educational subjects, such as English and mathematics, usually are included also. The Associate Degree is awarded and students who decide to go on for professional engineering training may transfer some of the credits they have earned in the two-year programs.

ARIZONA

Yavapai College, Prescott, AZ 86301. Associate Degree in science, solar energy technology.

ARKANSAS

Mississippi County Community College, Blytheville, AR 72315. Associate Degree in applied science, solar energy technology.

CALIFORNIA

Cabrillo College, Aptos, CA 95003. Associate Degree in science, solar technology.

Cerro Coso Community College, Ridgecrest, CA 93555. Associate Degree in applied science, solar engineering technology.

Cosumnes River College, Sacramento, CA 95823. Associate Degree in environmental design, solar energy systems.

Mount Saint Antonio College, Walnut, CA 91789. Associate Degree in air conditioning, heating, and ventilating (includes training in solar technology).

San Diego Community College, Evening College, San Diego, CA 92101. Associate Degree in air conditioning, heating, refrigeration, and solar technology.

San Jose City College, Solar Technology Department, 2100 Moor Park Avenue, San Jose, CA 95128. Associate Degree in science, solar technician.

COLORADO

Colorado Technical College, 655 Elkton Drive, Colorado Springs, CO 80907. Associate Degree in applied science, solar engineering technology.

Community College of Denver, Red Rock Campus, 12600 West 6th Avenue, Golden, CO 80401. Associate Degree in solar energy installation and maintenance.

Otero Junior College, La Junta, CO 81050. Associate Degree in applied science, solar heating option.

FLORIDA

Brevard Community College, Cocoa, FL 32922. Associate Degree in applied science, solar engineering technology.

Gulf Coast Community College, Panama City, FL 32401. Associate Degree in science, solar energy and solar systems.

Miami-Dade Community College, Miami, FL 33176. Associate Degree in science, air conditioning engineering technology.

Pensacola Junior College, Pensacola, FL 32504. Associate Degree in solar energy technology.

GEORGIA

DeKalb Community College, Clarkston, GA 30021. Associate Degree in solar heating.

---

* People already skilled in one of the trades will find opportunities for learning its solar aspects in the next section on "Continuing Education." Some of the technical programs listed there also are appropriate for people starting out in a trade, who want shorter, noncredit training.

ILLINOIS

Community College of Eastern Illinois, Olney Center Campus, Olney, IL 62450. Associate Degree in applied science, construction energy program.

IOWA

Scott Community College, Belmont Road, Bettendorf, IA 52722. Trade and Industrial Occupation Division gives Associate Degree in solar energetics technology.

Western Iowa Tech, Sioux City, IA 51102. Associate Degree in applied science, in solar systems technology.

KANSAS

Barton County Community College, Great Bend, KS 67530. Associate Degree in applied science, solar energy technology.

Kansas Technical Institute, Salina, KS 67401. Associate Degree in science, mechanical engineering technology–solar option.

MASSACHUSETTS

Cape Cod Community College, West Barnstable, MA 02668. Associate Degree in science, energy systems technology.

Springfield Technical Community College, Springfield, MA 01105. Associate Degree in engineering technology, solar energy option.

Franklin Institute, Boston, MA 02116. Associate Degree, includes intensive course in solar and alternative energy systems design.

MICHIGAN

Charles S. Mott Community College, Flint, MI 48503. Associate Degree in applied science, alternative energy.

Ferris State College, Big Rapids, MI 49307. Associate Degree in applied science in refrigerating, heating, and air conditioning (solar included as part of all regular courses).

Grand Rapids Junior College, Grand Rapids, MI 49502. Associate Degree in heating, ventilating, and air conditioning, or architectural drafting; solar courses part of program.

Lansing Community College, Lansing, MI 48901. Associate Degree in engineering technology.

MINNESOTA

Lakewood Community College, White Bear Lake, MN 55110. Associate Degree in applied science, energy engineering technology.

NEBRASKA

Metropolitan Technical Community College, Omaha, NE 68137. Associate Degree in solar systems.

NEVADA

Clark County Community College, Las Vegas, NV 89030. Associate Degree in applied science, solar energy technology.

NEW YORK

Mohawk Valley Community College, 1101 Sherman Drive, Utica, NY 13503. Associate Degree in solar energy technology.

SUNY Agricultural and Technical College, Delhi, NY 13753. Associate Degree in applied science, construction technology.

NORTH CAROLINA

Cape Fear Technical Institute, Wilmington, NC 28401. Associate Degree in general occupational technologies.

OREGON

Linn-Benton Community College, Albany, OR 97321. Associate Degree in engineering technology, solar energy option.

PENNSYLVANIA

Keystone Junior College, La Plume, PA 18440. Associate Degree in applied science, solar engineering technology.

Lehigh County Community College, Schnecksville, PA 18078. Associate Degree in applied science, alternative energy technologies.

Pennsylvania Institute of Technology, 414 Sansom Street, Upper Darby, PA 19082. Associate Degree in energy technology.

Triangle Institute of Technology, Inc., 635 Smithfield Street, Pittsburgh, PA 15222. Associate Degree in specialized technology.

RHODE ISLAND

New England Institute of Technology, 184 Early Street, Providence, RI 02907. Associate Degree in refrigeration, air conditioning and heating.

SOUTH CAROLINA

Beaufort Technical Education Center, Beaufort, SC 29902. Associate Degree in refrigeration and air conditioning, with intensive course in solar energy application.

TENNESSEE

Motlow State Community College, Tullahoma, TN 37388. Associate Degree in energy, engineering technology.

TEXAS

Central Texas College, Killeen, TX 76541. Associate Degree in applied science, solar energy, systems technology.

Navarro College, Corsicana, TX 75110. Associate Degree in applied science, solar engineering technology. Navarro did an extensive study of the future demand for solar technicians, the skills and type of training needed for design, installation, and maintenance of solar systems. The study serves as a resource for other two-year solar technology programs.

North Lake College, Irving, TX 75062. Associate Degree in solar technology.

Ranger Junior College, Ranger, TX 76470. Associate Degree in applied science, air conditioning, and refrigeration–solar energy option.

WYOMING

Sheridan College, Sheridan, WY 82801. Associate Degree in engineering technology, solar option.

Continuing Education for Solar Technologies

The educational opportunities listed in this section are for people who want to increase their technical knowledge of solar energy and energy conservation but are not concerned with academic credit. Researchers, engineers, architects, building contractors, and skilled tradespeople will find here some of the places where they can learn to turn their occupations toward solar energy, or keep up-to-date in technical solar developments.

Continuing education is often in the form of short-term seminars, workshops, or conferences, but some of the courses run over a period of weeks on a regular class schedule. A few intensive training programs run for several weeks, full time. If offered locally, continuing education programs are listed by state.

National Organizations and Institutions

The first listings are organizations and institutions that provide seminars, workshops, and conferences periodically at several different locations throughout the United States and Canada.

American Institute of Architects, 1735 New York Avenue, Washington, DC 20006.

American Institute of Chemical Engineers, 345 East 42nd Street, New York, NY 10017.

American Society of Heating, Refrigerating, and Air Conditioning Engineers, 345 East 47th Street, New York, NY 10017.

American Society of Mechanical Engineers, 345 East 47th Street, New York, NY 10017.

American Section, International Solar Energy Society, American Technological University, P.O. Box 1416, Killeen, TX 76541. The regional and local chapters also have educational programs (see Part IV, "Organizations").

Engineering Foundation, 345 East 47th Street, New York, NY 10017.

Government Institutes, Inc., 4733 Bethesda Avenue NW, Washington, DC 20014.

Information Transfer, Inc., 1160 Rockville Pike, Rockville, MD 20852.

Institute of Electrical and Electronic Engineers, 345 East 47th Street, New York, NY 10017.

Institute of Gas Technology, 3424 South State Street, Chicago, IL 60616.

Jordan College, Energy Division, 360 West Pine Street, Cedar Springs, MI 49319.

New York University, School of Continuing Education, Division of Career Development, 326 Simkin Hall, New York, NY 10003.

North American Heating and Air Conditioning Wholesalers Association, 1161 West Henderson Road, Columbus, OH 43220. Home study course in the installation and maintenance of solar heating and cooling.

Regional Solar Development Centers

Mid-American Solar Energy Complex, 1256 Trapp Road, Eagan, MN 55121.

Northeast Solar Energy Center, 70 Memorial Drive, Cambridge, MA 02142.

Southern Solar Energy Center, Exchange Place, Suite 1250, 2300 Peachford Road, Atlanta, GA 30338.

Western SUN, Portland, OR 97208.

Solar Energy Industries Association, 1001 Connecticut Avenue NW, Suite 632, Washington, DC 20036.

Solar Energy Research Institute, 1536 Cole Boulevard, Golden, CO 80401.

Solar Energy Society of Canada, Inc., 608-870 Cambridge Street, Winnipeg, Manitoba, Canada R3M 3H5.

Trinity University, Continuing Education Division, 715 Stadium Drive, San Antonio, TX 78264.

U.S. Department of Labor, Office of Comprehensive Employment Development Programs, 200 Constitution Avenue NW, Washington, DC 20210. Provides CETA (Comprehensive Employment Training Act) employees for local and regional projects that have training programs in weatherizing and in passive solar construction.

By State

ALABAMA

University of Alabama, Department of Mechanical Engineering, Huntsville, AL 35807. Conferences on applications of solar energy.

ARKANSAS

University of Central Arkansas, Conway, AR 72032. Twelve-hour solar energy workshop for contractors, plumbers, and sheet metal workers; awards certificate of completion.

CALIFORNIA

University of California at Los Angeles, Los Angeles, CA 90024. Department of Continuing Education in Engineering and Mathematics uses the structure and facilities of the statewide University of California Extension organization to provide an extensive program of evening classes, concentrated short courses, and conferences in all university locations.

Center for Employment Training, 425 South Market Street, San Jose, CA 95113. Two weeks of training in building and installing solar collectors as part of a six-week (full-time) course in building maintenance.

Chaffey Community College, 5885 Haven Avenue, Alta Loma, CA 91701. Two courses in installation of solar systems.

Coastline Community College, Fountain Valley, CA 92708. Ten courses in energy management for solar technicians.

Energy Systems, Inc., 4570 Alvarado Canyon Road, San Diego, CA 92120. Hands-on solar energy training course for dealers and installers.

Solarcon, P.O. Box 14875, San Francisco, CA 94114. Workshop covering solar systems installation.

Solar Technician Training Program, Office of Appropriate Technology, 1322 O Street, Sacramento, CA 95814. Six-month job training program to gain working skills in installing solar hot water systems.

COLORADO

Colorado Office of Energy Conservation, 1410 Grant Street, B-104, Denver, CO 80203. Runs series of workshops for builders, contractors, engineers, architects, and planners.

Colorado State University, Solar Energy Applications Laboratory, Fort Collins, CO 80523. Three- to five-day seminars on solar systems design.

Domestic Technology Institute, 12520 West Cedar Drive, Lakewood, CO 80228. Workshops on solar energy and conferences on community technology.

Solar Energy Research Institute, 1536 Cole Boulevard, Golden, CO 80401.

University of Colorado, Center for Management and Technical Programs, Boulder, CO 80309. Two-day seminars on commercial and industrial applications of solar energy.

CONNECTICUT

The Hartford Graduate Center, 275 Windsor Street, Hartford, CT 06120. Solar Energy Seminars Program covers design, installation, maintenance, monitoring, and performance prediction of solar heating and hot water systems; offered in evening courses or in one- to five-day seminars.

DELAWARE

Newcastle County Vocational-Technical School, 1417 Newport Road, Wilmington, DE 19804. One course in installing solar heating in buildings.

University of Delaware, Continuing Professional Education, Newark, DE 19711. Five-day intensive seminar on design of residential solar systems.

DISTRICT OF COLUMBIA

George Washington University, Continuing Engineering Education Program, Washington, DC 20052. Five-day (full-time) courses on advances in photovoltaics and on residential and commercial applications of solar heating and cooling.

FLORIDA

South Florida Technical Institute, Department of Air Conditioning, Refrigeration, and Major Appliances, 201 West Sunrise Boulevard, Fort Lauderdale, FL 33311. Five-week (full-time) course in installing solar systems.

University of Miami, Coral Gables, FL 33124. School of Continuing Studies sponsors conferences and technical meetings on alternative energy sources.

GEORGIA

DeKalb Community College, Clarkston, GA 30021. Fourteen-week (full-time) course in installing solar systems in Department of Heating and Air Conditioning.

Georgia Institute of Technology, Continuing Education Division, Atlanta, GA 30332. Solar courses for engineers and scientists.

ILLINOIS

Illinois Eastern Lincoln Trail College, Robinson, IL 62454. One course in installing solar systems in Department of Air Conditioning and Refrigeration.

Institute of Gas Technology, 3424 South State Street, Chicago, IL 60616. Three-day workshops on technologies for energy storage, energy modeling, calculating net energy gain, design, installation, and operation criteria for solar systems.

Quincy Technical School, Quincy, IL 62301. One-week (full-time) course on fundamentals of solar space and water heating in Department of Air Conditioning, Refrigeration, and Heating Service.

Solar Store, Inc., Box 841, Peoria, IL 61652. Solar energy workshops for installers.

INDIANA

Purdue University, Continuing Engineering Education, Fort Wayne, IN 46805. One course in solar principles and design of systems.

MAINE

Portland Vocational Center, Portland, ME 04111. One course in solar systems installation.

MARYLAND

Rets Technical Center, 511 Russell Street, Baltimore, MD 21230. Six-week (full-time) course in installing solar systems.

MASSACHUSETTS

Blue Hills Regional Technical Institute, 100 Randolph Street, Canton, MA 02172. One course in solar heating systems design in Department of Heating, Ventilating, and Air Conditioning.

Boston Architectural Center, 320 Newbury Street, Boston, MA 02115. Service for Energy Conservation in Architecture (SECA), a nonprofit organization within the center, runs forums and workshops on energy conservation in commercial buildings and on building codes that affect energy use. It has developed an eight-course Energy Curriculum for Continuing Education.

MIT Laboratory of Architecture and Planning, 77 Massachusetts Avenue, Room 4209, Cambridge, MA 02139. Three-day seminars for architects and engineers.

New England Fuel Institute, Solar Education Division, 20 Summer Street, P.O. Box 888, Watertown, MA 02172. Four-week (full-time) hands-on training for installing and maintaining solar heating and hot water systems.

Northeast Institute of Industrial Technology, 41 Phillips Street, Boston, MA 02114. One course in installing solar hot water heaters in Department of Air Conditioning and Refrigeration Technology.

MICHIGAN

General Motors Institute, Continuing Engineering Education, Flint, MI 48502. One-week (full-time) course in solar energy for engineers, architects, and educators.

Jordan College, Energy Division, 360 West Pine Street, Cedar Springs, MI 49319. Semester and short-term technical courses, seminars, and conferences on solar, wind, geothermal, hydroelectric, and biomass.

University of Michigan, Continuing Engineering Education, 300 Chrysler Center, North Campus, Ann Arbor, MI 48109. Four-day (full-time) workshops for architects and engineers on specific solar topics, such as measurements and instrumentation.

MONTANA

Montana State University, Bozeman, MT 59715. One course on energy-efficient building for contractors.

NEVADA

University of Nevada at Las Vegas, Las Vegas, NV 89154. Courses in solar systems, including heating, cooling, hot water and swimming pools, and in the economics of solar systems, run by the Continuing Education Division and the Nevada Solar Energy Association.

NEW HAMPSHIRE

Total Environmental Action, Harrisville, NH 03450. Advanced workshops on passive solar design.

New Hampshire Vocational-Technical Institute, 1066 Front Street, Manchester, NH 03102. Courses in solar installations; solar house project.

NEW JERSEY

Essex County Technical Careers Center, 91 West Market Street, Newark, NJ. Day and evening courses in installing solar heating systems; awards certificate of completion.

Mercer County Area Vocational Technical Schools, 1085 Old Trenton Road, Trenton, NJ 08690. One course on installing solar systems in Department of Plumbing, Heating, and Refrigeration.

Ocean County Vocational Technical Schools, Route 571, Jackson, NJ 08527. Evening school courses in solar systems installation in Department of Heating, Ventilating, and Air Conditioning; awards certificate of completion.

Passaic School of Drafting, 657 Main Avenue, Passaic, NJ 07055. Four-week (full-time) course in architectural drafting, solar energy design.

Salem County Vocational Technical Schools, RD 2, Box 350, Woodstown, NJ 08098. Courses on solar energy for plumbing and heating trades.

Southern New Jersey OIC, Camden, NJ. One course in solar unit installation; awards certificate of completion.

Union County Technical Institute, Scotch Plains, NJ. One course in installing solar systems in Department of Heating, Ventilating, and Air Conditioning.

NEW MEXICO

New Mexico Solar Energy Institute, New Mexico State University, Las Cruces, NM

88003. Seminars, short courses in solar building and other technical topics for architects, designers, builders; has information about other solar courses in the state.

University of New Mexico, Technology Applications Center, Albuquerque, NM 87131. Three-day seminars in solar energy systems design and applications.

NEW YORK

American Institute of Chemical Engineers, 345 East 47th Street, New York, NY 10017. Two-day seminars on solar energy applications and process design for energy conservation; also given in other cities.

American Society of Mechanical Engineers, 345 East 47th Street, New York, NY 10017. Seminars and short courses for professional development in solar energy; also given in other cities.

Adirondack Community College, Glens Falls, NY 12801. One course in installing solar systems; awards certificate of completion.

New York University, School of Continuing Education, Division of Career Development, 326 Simkin Hall, New York, NY 10003. One-week courses covering principles and applications of most solar technologies; three-day seminars for professionals given in major cities.

State University of New York at Albany, Albany, NY 12222. Atmospheric Sciences Research Center runs one-week solar energy workshop.

NORTH CAROLINA

Carteret Technical Institute, Continuing Education, Morehead City, NC 28557. One course in solar energy fundamentals and construction; awards certificate of completion.

Coastal Carolina Community College, Jacksonville, NC 28540. One course in installing solar systems in Department of Heating, Air Conditioning, and Refrigeration.

Pamlico Technical Institute, Continuing Education, Grantsboro, NC 28529. One course in installing solar systems; awards certificate of completion.

Southwestern Technical Institute, Sylva, NC 28779. Program of several courses in solar energy systems—residential and commercial construction; awards certificate of completion.

Technical Institute of Almamance, Haw River, NC 27258. One course in installing solar systems in Department of Air Conditioning and Refrigeration.

NORTH DAKOTA

Bismarck Junior College, Bismarck, ND 58501. Eight-week (full-time) course in installing solar systems in Department of Heating, Refrigeration, and Air Conditioning; awards certificate of completion.

North Dakota State School of Science, Wahpeton, ND 58075. One course in installing solar systems in Department of Environmental Studies; awards certificate of completion.

OHIO

Kent State University, Main Campus, Continuing Education, Kent, OH 44242. Architects and energy course given by Department of Architecture.

University of Dayton, Research Institute, 300 College Park Avenue, Dayton, OH 45469. Technical workshops in solar heating design; provides speakers on solar energy for general public.

OKLAHOMA

Oklahoma State University, Main Campus, Stillwater, OK 74074. Architecture Extension and Technology Extension run a number of one-week courses in alternative energy systems, energy-conserving design, design of solar heating systems.

OREGON

University of Oregon, Solar Energy Center, Eugene, OR 97403. One-week summer workshops in solar monitoring and data measurement (in conjunction with Oregon State University Department of Atmospheric Sciences) and other short seminars on solar topics.

PENNSYLVANIA

Drexel University, Office of Continuing Professional Education, 32nd & Chestnut Streets, Philadelphia, PA 19104. Evening courses in solar heating and cooling systems.

Northampton County Area Community College, Bethlehem, PA 18017. Course on design and utilization of emerging energy sources, with 60 hours of laboratory work, in Vocational-Technical Department.

Pennsylvania State University, Allentown Campus, Fogelsville, PA 18051. Workshops in solar space heating and cooling, hot water systems.

Pennsylvania State University, Shenango Valley Campus, Sharon, PA 16146. Two-week (full-time) course in solar heating and cooling in Department of Physics; awards short course certificate.

Wilkes College, Department of Engineering, Wilkes-Barre, PA 18703. Three-week (full-time) professional course in energy conscious alternatives.

SOUTH CAROLINA

Florence Darlington Tech, Florence, SC 29502. One course in installing solar systems in Department of Industrial Trades.

Tri-County Technical College, Pendleton, SC 29670. One course on solar energy applications in Department of Air Conditioning and Refrigeration.

Trident Technical College, P.O. Box 10367, Charleston, SC 29411. Three-week (full-time) course in installing solar systems in Department of Air Conditioning and Refrigeration.

York Technical College, Rock Hill, SC 29730. One course in installing solar systems in Department of Air Conditioning, Refrigeration, and Heating.

TENNESSEE

Cleveland State Community College, Cleveland, TN 37311. Two-week (full-time) faculty development workshop in energy alternatives.

University of Tennessee, College of Engineering, Knoxville, TN 37916. Four-day seminars on heating and cooling of buildings; sponsors solar utilization conferences.

TEXAS

Central Texas College, Killeen, TX 76541. Several courses in installing solar systems in Department of Industrial Technology; awards certificate of completion.

Odessa College, Odessa, TX 79760. One course in installing solar systems in Department of Refrigeration and Air Conditioning.

Texas A & M University, Office of Continuing Education, College Station, TX 77843. One-week (full-time) course in solar heating and cooling and energy-conscious design in buildings; also offers two-day Applied Solar Energy Seminars cosponsored by Department of Mechanical Engineering and Solar Engineering.

Texas State Technical Institute, Sweetwater, TX 79556. One course in installing solar systems in Department of Air Conditioning and Refrigeration Technology.

Texas State Technical Rio Grande, Harlingen, TX 78550. One course in installing solar systems in Department of Air Conditioning and Refrigeration Technology; awards certificate of completion.

Trinity University, Continuing Education, 715 Stadium Drive, San Antonio, TX 78264. Courses and seminars in solar systems principles and design.

University of Houston, Central Campus, Houston, TX 77004. Energy Laboratory runs one- and two-day technical workshops and weekly afternoon seminars covering energy conservation and solar technologies.

UTAH

Dixie College, Saint George, UT 84770. Two courses in installing solar systems in Department of Engineering Technology; awards certificate of completion.

WASHINGTON

North Seattle Community College, Seattle, WA 98103. Two courses in installing solar systems in Department of Engineering Technology.

WISCONSIN

Moraine Park Technical Institute, Fond du Lac, WI 54935. Several courses in installing solar systems, one in wind energy, in Trade and Technical Department.

District One Technical Institute, Eau Claire, WI 54701. One course in alternative energy systems in Department of Air Conditioning Technology.

University of Wisconsin, Extension Department of Engineering and Applied Science, 432 North Lake Street, Madison, WI 53706. A number of one- to four-day (full-time) technical courses in solar heating and cooling of buildings, solar energy thermal processes, earth-sheltered architecture, energy management, and residential energy auditing.

University of Wisconsin at La Crosse, La Crosse, WI 54601. One-week (full-time) course in alternative energy sources and principles of solar energy, run by Department of Physics.

University of Wisconsin at Oshkosh, Oshkosh, WI 54901. Three-week (full-time) course in technologies of solar energy, run by Department of Physics.

University of Wisconsin at Whitewater, Whitewater, WI 53190. Two-week (full-time) course in residential solar heating, run by Department of Physics.

Wisconsin Indian Vocational, Technical, and Adult Education, 600 North 21st Street, Superior, WI 54880. One course in installing solar systems in Department of Facilities Engineering.

Canadian

ALBERTA

University of Alberta, Faculty of Extension, Corbett Hall, Edmonton, Alberta T6G 2G4. Technical courses in design of buildings using conservation techniques and solar heating systems.

University of Calgary, Calgary, Alberta T2N 1N4. Course in hand calculation and computer simulation for architects and engineers who have basic knowledge of solar principles.

BRITISH COLUMBIA

University of British Columbia, Centre for Continuing Education, Vancouver, British Columbia V6T 1W5. Intensive course in energy conservation and solar energy design for architects and engineers; three-day conference for teachers of grades K–12, to develop energy education.

MANITOBA

University of Manitoba, Winnipeg, Manitoba R3T 3C8. One-week hands-on course on construction of energy-efficient housing and use of solar heating systems; course for college graduates in the design and evaluation of solar heating systems.

NEW BRUNSWICK

University of New Brunswick, Department of Extension, Box 4400, Fredericton, New Brunswick E3B 5A3. Short state-of-the-art course in solar energy for professional engineers and architects; solar conferences.

ONTARIO

Brock University, Office of Part-time Programs, St. Catherines, Ontario L2S 3A1. Four-day seminar in solar technologies and Canadian solar projects.

Lambton College, P.O. Box 969, Sarnia, Ontario N7T 7K4. Courses deal with design and detail of solar systems, plus pool heating and greenhouses; one course covers other alternative technologies, especially wood, wind, and biogas.

Ryerson Polytechnical Institute, Continuing Education Division, 50 Gould Street, Toronto M5B 1E8. Series of courses on construction and operation of solar energy heating systems.

University of Guelph, Office of Continuing Education, 145 Johnson Hall, Guelph, Ontario N1G 2W1. Four-day workshop on solar heating design and materials selection.

QUEBEC

Ecole Polytechnique, Montreal, Quebec H3C 3A7. Three-day course for architects and engineers in solar utilization in buildings.

SASKATCHEWAN

University of Saskatchewan, Saskatoon, Saskatchewan S7N 0W0. Course on energy conservation and passive techniques in building design; short hands-on course in construction of collectors for solar water heating systems.

General Solar Education

Special College Programs

There are several special college programs for those who want a broad exposure to the technical and nontechnical aspects of solar energy. Some students in these programs go on for further training in a particular science or technology; some want to continue working on the psychological, social, or economic implications of solar energy; others find jobs following graduation using their knowledge of solar energy and their existing skills. Several of the new college programs are described below.

Antioch College-West, 1161 Mission Street, San Francisco, CA 94103.
A BA or an MS may be earned in the Environmental Studies Program with a major in solar energy and design, appropriate technology, ecosystem management, or other alternative energy systems. With the aid of a faculty committee, students design individual programs of study that include the conceptual background, methodology, and applied skills needed in the chosen field. Classroom work is combined with independent study, tutorials, group workshops, field work, and internships with outside experts. Academic credit is given for prior learning, gained both in and out of school, if properly documented, and also for job-related study while pursuing a degree.

Goddard College, Plainfield, VT 05667.
Social Ecology Program leads to an MA degree. It involves multidisciplinary study, in combination with specialized knowledge and skills in one field. Specialized includes solar and wind energy and energy-efficient shelter design and construction. Students enter the program in June for a 12-week resident term, then participate in an internship, and, from September through May, complete an individual study and research plan. Goddard's Cate Farm offers opportunities for research and study of the experimental solar systems installed there. Installations include greenhouses, windmills, solar collectors, solar water heaters, and aquaculture facilities.

Governor's State University, Park Forest South, IL 60466.
Graduate and undergraduate degree candidates in environmental sciences may earn credit by working on *Outlook*, the university's

appropriate technology publication. The *Outlook* staff also produce films and TV programs, maintain a speakers bureau and reference collection, and help to coordinate conferences. The emphasis is on appropriate technologies, energy alternatives, and community policy and planning.

Jordan College, Energy Division, 360 West Pine Street, Cedar Springs, MI 49319.

A small liberal arts college, with a belief in combining theory and practice, Jordan started in the alternative energy field with low-technology solar devices made and installed by students and faculty. It now has seven alternative energy units on the main campus, which serve as demonstration sites and greatly reduce energy costs. Courses on alternative energy were added to the curriculum, mini-courses offered to the public, funding obtained from the U.S. Department of Energy for solar installations, until the College became an important resource for the region as a provider of alternative energy demonstration and teaching facilities.

The *Jordan Energy Institute* is the locale for a core curriculum on five alternative energy systems: solar, wind, geothermal, hydroelectric, and biomass, taught in the classroom and laboratory and with hands-on experience where possible. The college encourages simple patterns of living, conservation, and do-it-yourself resourcefulness. Most of the buildings are constructed of recycled wood from barns and were erected with the help of faculty and students. Degrees offered: BS in alternative energy-environmental studies; Associate Degree in applied science in alternative energy. Two-year students may transfer for full engineering degree.

Marlboro College, Marlboro, VT 05344.

Programs in alternative energy started in 1974 when eight students made comprehensive plans for converting the entire campus to energy self-sufficiency. Students are involved in implementing the plan, which will make the campus a working model for small, energy-conscious New England communities. To date they have converted a college residential house to a self-sufficient system, using a combination of flat plate collectors and a wood-burning furnace; constructed an active and passive solar- and wood-heated residence in the town; designed and built two solar-heated greenhouses attached to the science building, where they monitor agricultural performance and amount of heat supplied to the building; participated in a national solar greenhouse conference held on campus and in other campus alternative energy conferences.

Students, with faculty advice and approval, design their own plans of concentration, including outside internships and a major senior year project. Special courses, such as one on Solar Energy and Building Design, are combined with courses in liberal arts and sciences. A BS in alternative energy studies is awarded.

Washington University, School of Engineering and Applied Science, Box 1164, St. Louis, MO 63130.

BS or MA degree programs in technology and human affairs enable students with strong interests and aptitudes in science, mathematics, and engineering to develop the qualitative and quantitative skills needed to solve social problems involving both technology and public policy. The program provides a liberal-technical or preprofessional education and places more emphasis on the humanities and social science courses than does the university's Engineering Program. Students are encouraged to plan programs in relation to their particular interests. Many students and faculty work on projects or policy studies in cooperation with outside government and private agencies. Available at the university are research laboratories and a well-developed solar test facility, as well as several courses in the solar field, for instance, Solar Energy Technology and Policy. Students in the Technology and Human Affairs Department have formed an organization of their own.

West Virginia University, Program for the Study of Technology, Suite 609 Allen Hall, Morgantown, WV 26506.

This program leads to an MA in appropriate technology. It explores the technical, social-cultural, and historical aspects of appropriate technology, and also deals with the contemporary problems and issues related to different types of technologies. Current interests and investigations include shelter, energy, food, resources, manufacturing, communication, transportation, philosophy and values, alternative life styles, community development, social control, survival skills, self-sufficiency, developing nations, technology transfer, and technology assessment. Students participate in formal course work, seminars, and independent and group study, seeking an integration of the technical and social-cultural aspects of appropriate technology.

Continuing Education*

Conferences, short-term seminars, workshops, and individual courses in solar energy give people considering solar jobs, and other interested citizens, an introduction to the principles and techniques of solar energy. General solar education is offered in university continuing education divisions and in adult programs at high schools and private educational institutions. It is also available at community colleges, many of which include one or more solar courses in their curricula. Community colleges usually allow people to take individual courses without becoming degree candidates.

United States

Alabama

University of Alabama, Huntsville, AL 35807
Auburn University, Montgomery, AL 36117

Alaska

Anchorage Community College, Anchorage, AK 99504
Kenai Community College, Soldothna, AK 99669
Northwest Community College, Nome, AK 96762
Tanana Valley Community College, Fairbanks, AK 99701

Arizona

Cochise College, Douglas, AZ 85607
Glendale Community College, Glendale, AZ 85302
Mohave Community College, Kingman, AZ 86401
Northland Pioneer College, Holbrook, AZ 86025
Rio Salado Community College, 10451 Palmero Drive, Sun City, AZ 85313

California

University of California at Davis, Davis, CA 95616.

The Energy Extension Service here is the lead group within the statewide university system for energy education and information. It develops and coordinates energy courses through the extension program and acts as a clearinghouse for information related to all types of energy education throughout California. There are evening classes, concentrated short courses, and conferences at many university locations.

American River College, Sacramento, CA 95841
Bakersfield College, Bakersfield, CA 93305
Butte College, Oroville, CA 95965
Chaffey College, Alta Loma, CA 91701
Citrus College, Azusa, CA 91702
College of the Desert, Palm Desert, CA 92260
College of Marin, Kentfield, CA 94904
College of the Redwoods, Eureka, CA 95501
College of the Sequois, Visalia, CA 93277

---

* "Organizations" listings carry other resources for general solar continuing education; people who want a more rigorous technical content should refer to "Continuing Education for Solar Technologies," p. 192.

College of the Siskiyous, Weed, CA 96094
Cuesta College, San Luis Obispo, CA 93406
De Anza College, Cupertino, CA 95014
Diablo Valley College, Pleasant Hill, CA 94523
Evergreen Valley College, San Jose, CA 95121
Feather River College, Quincy, CA 95971
Foothill College, Los Altos Hills, CA 94022
Fresno City College, Fresno, CA 93704
Fullerton College, Fullerton, CA 92634
Gavilan College, Gilroy, CA 95020
Glendale Community College, Glendale, CA 91208
Los Angeles Pierce College, Woodland Hills, CA 91371
Los Angeles Trade Technical College, Los Angeles, CA 90015
Merced College, Merced, CA 95340
Modesto Junior College, Modesto, CA 95350
Moorpark College, Moorpark, CA 93021
Mt. San Jacinto College, San Jacinto, CA 92383
Napa College, Napa, CA 94558
Ohlone College, Fremont, CA 94537
Orange Coast College, Costa Mesa, CA 92626
Pasadena City College, Pasadena, CA 91106
Riverside City College, Riverside, CA 92506
Sacramento City College, Sacramento, CA 95822
San Bernadino Valley College, San Bernadino, CA 92403
San Diego Community College, Mesa College, San Diego, CA 92111
San Joaquin Delta College, Stockton, CA 95207
Santa Ana College, Santa Ana, CA 92706
Sierra College, Rocklin, CA 95677
West Valley College, Saratoga, CA 95070

Colorado

Colorado Mountain College, West Campus, Glenwood Springs, CO 81601
Community College of Denver, North Campus, Westminster, CO 80030
Domestic Technology Institute, Box 2043, Evergreen, CO 80439
Mesa College, Grand Junction, CO 81501
Solar Energy Research Institute, 1536 Cole Boulevard, Golden, CO 80401
Trinidad State Junior College, Trinidad, CO 81082
University of Colorado, Educational Media Center, Stadium 360, Boulder, CO 80309
University of Colorado, Division of Continuing Education, Colorado Springs, CO 80907
University of Colorado at Denver, 1100 14th Street, Denver, CO 80202
University of Denver, Denver, CO 80201

Connecticut

Energy Education Services of Connecticut, P.O. Box 224. Hartford, CT 06103
Thames Valley State Technical College, Norwich, CT 06360

District of Columbia

George Washington University, Washington, DC 20052

Florida

Edison Community College, Fort Myers, FL 33901
Florida Solar Energy Center, 300 State Road, Cape Canaveral, FL 32920
Pinellas Vo-Tech Institute, 6100 154th Avenue N., Clearwater, FL 33516
Santa Fe Community College, Gainesville, FL 32601
University of Miami, Coral Gables, FL 33124
Valencia Community College, Orlando, FL 32802

Georgia

Georgia Institute of Technology, Continuing Education, Atlanta, GA 30332

Hawaii

University of Hawaii, Honolulu Community College, Honolulu, HI 96766
University of Hawaii, Kauai Community College, Lihue, HI 96766

Idaho

University of Idaho, Moscow, ID 83843

Illinois

College of Dupage, Glen Ellyn, IL 60137
John A. Logan College, Carterville, IL 62918
Kankakee Community College, Kankakee, IL 60901
McHenry County College, Crystal Lake, IL 60014
Moraine Valley Community College, Palos Hills, IL 60465
Sangamon State University, Springfield, IL 62703
Triton College, River Grove, IL 60171
Wm. Rainey Harper College, Palatine, IL 60067

Indiana

Indiana Vocational Technical College, Evansville, IN 47710, Sellersburg, IN 47172, Gary, IN 46409

Iowa

Des Moines Area Community College, Ankeny, IA 50021
Kirkwood Community College, Cedar Rapids, IA 52406
Muscatine Community College, Muscatine, IA 52761

Kansas

Garden City Community College, Garden City, KS 67846
Kansas State University, Center for Energy Studies, Manhattan, KS 66502
University for Man, 1221 Thurston Avenue, Manhattan, KS 66502

Louisiana

Delgado College, New Orleans, LA 70119

Maine

Cornerstones, Wing School of Shelter Technology, 54 Cumberland Street, Brunswick, ME 04011
Shelter Institute, 58 Center Street, Bath, ME 04530
University of Southern Maine, 96 Falmouth Street, Portland, ME 04103

Maryland

Allegheny Community College, Cumberland, MD 21502
Dundalk Community College, Baltimore, MD 21222
Harford Community College, Bel Air, MD 21014

Massachusetts

Boston University, Center for Energy Studies, Boston, MA 02215
Bristol Community College, Fall River, MA 02722
Bunker Hill Community College, Charlestown, MA 02129
The Cambridge School, Weston Center for Open Education, Weston, MA 02193
Habitat Institute, 10 Juniper Road, Box 136, Belmont, MA 02178
Heartwood Owner-Builder School, Washington, MA 01235
Hoosuck Institute, Windsor Mill, North Adams, MA 01247
New Alchemy Institute, P.O. Box 432, Woods Hole, MA 02543
North Shore Community College, Beverly, MA 01915

Michigan

Delta College, University Center, MI 48710
Glen Oaks Community College, Centreville, MI 49032
Jackson Community College, Jackson, MI 49201
Jordan College, 360 West Pine Street, Cedar Springs, MI 49319
Macomb County Community College, South Campus, Warren, MI 48093
Mid-Michigan Community College, Harrison, MI 48625
Schoolcraft College, Livonia, MI 48152
Saint Clair County Community College, Fort Huron, MI 48060
Southwestern Michigan College, Dowagiac, MI 49047
Sun Structures, 201 East Liberty Street, Ann Arbor, MI 48109

Wayne State University, College of Lifelong Learning, Detroit, MI 48202

Minnesota

Rochester Community College, Rochester, MN 55901

Mississippi

Hinds Junior College, Raymonds, MS 39154

Missouri

Crowder College, Neosho, MO 64850
Saint Louis Community College-Meramec, Kirkwood, MO 63122
University of Missouri, Continuing Education, Columbia, MO 65201

Montana

Alternative Energy Resource Organization, 435 Stapleton Building, Billings, MT 59101
Flathead Valley Community College, Kalispell, MT 59901

Nebraska

Central Technical Community College, Grand Island, NE 68801
Midland Lutheran College, Fremont, NE 68025
Mid-Plains Community College, North Platte, NE 69101
Southeast Community College, Milford, NE 68405

Nevada

University of Nevada at Las Vegas, Continuing Education-Nevada Solar Energy Association, Las Vegas, NV 89154

New Hampshire

New England Center for Appropriate Technology, 15 Garrison Avenue, Durham, NH 03824
New Hampshire Vocational-Technical College, Manchester, NH 03101
New Hampshire Vocational-Technical College, Nashua, NH 03060
The Phoenix Nest, South Acworth, NH 03607
Total Environmental Action, Hamville, NH 03450

New Jersey

Brookdale Community College, Lincroft, NJ 07738
Middlesex County College, Edison, NJ 08817
Ocean County Vocational-Technical Schools, Route 571, Jackson, NJ 08527
Salem County Vocational Technical Schools, RD 2, Box 350, Woodstown, NJ 08098

New Mexico

College of Santa Fe, Center for Continuing Education, Santa Fe, NM 87501
New Mexico State University, Solar Energy Institute, Las Cruces, NM 88003
Northern New Mexico Community College, El Rito, NM 87530
University of New Mexico, Technology Applications Center, Albuquerque, NM 87131

New York

Cayuga County Community College, Auburn, NY 13021
Columbia-Greene Community College, Hudson, NY 12534
Genesee Community College, Batavia, NY 14020
Monroe Community College, Rochester, NY 14623
New York University, New York, NY 10012
Orange County Community College, Middletown, NY 10940
Rockland Community College, Suffern, NY 10901
State University of New York at Albany, Albany, NY 13021
State University of New York, Agricultural and Technical College, Canton, NY 13617
Tompkins-Cortland Community College, Dryden, NY 13053
Westchester Community College, Valhalla, NY 10595

North Carolina

Center Piedmont Community College, Charlotte, NC 28204

Caston College, Dallas, NC 28034
Guilford Technical Institute, Jamestown, NC 27282
Randolph Technical Institute, Asheville, NC 27203
Stanley Technical Institute, Albemarle, NC 28001
Tri-County Technical Institute, Murphy, NC 28906

Ohio

Central Ohio Technical College, Newark, OH 43055
Columbus Technical Institute, Columbus, OH 43216
Lakeland Community College, Mentor, OH 44060
Muskingum Area Technical College, Zanesville, OH 43701
Northwest Technical College, Archbold, OH 43502
Sinclair Community College, Dayton, OH 45402

Oregon

Clackamas Community College, Oregon City, OR 97045
Rogue Community College, Grant's Pass, OR 97526
University of Oregon, Solar Energy Center, Eugene, OR 97403

Pennsylvania

Bucks County Community College, Newtown, PA 18940
Delaware County Community College, Media, PA 19063
Drexel University, 32nd and Chestnut Streets, Philadelphia, PA 19104
Harrisburg Area Community College, Harrisburg, PA 17110
The School of Living, P.O. Box 3233, York, PA 17402

South Carolina

Chesterfield-Marboro Tech, Cheraw, SC 29520
Piedmont Technical College, Greenwood, SC 29646
Spartanburg Technical College, Spartanburg, SC 29303
Tri-County Technical College, Pendleton, SC 29670

Texas

Grayson County Junior College, Denison, TX 75020
Lee College, Baytown, TX 77520
Our Lady of Lake University, San Antonio, TX 78285
Tri-County Continuing Education, Abilene, TX 79604

Utah

Utah Technical College, Salt Lake City, UT 84107

Vermont

Community College of Vermont, Montpelier, VT 05602
St. Michael's College, Winooski, VT 05404

Virginia

Blue Ridge Community College, Weyers Cave, VA 24486
Danville Community College, Danville, VA 24541
Lord Fairfax Community College, Middletown, VA 22645
Northern Virginia Community College, Annandale, VA 22003
Thomas Nelson Community College, Hampton, VA 23670
Virginia Western Community College, Roanoke, VA 24015

Washington

Fort Steilacoom Community College, Tacoma, WA 98498
Olympia Technical Community College, Olympia, WA 98502
Peninsula College, Port Angeles, WA 98362
Spokane Falls Community College, Spokane, WA 99204
Tacoma Community College, Tacoma, WA 98465

West Virginia

Parkersburg Community College, Parkersburg, WV 26101

Potomac State College, Keyser, WV 26726

Wisconsin

Milwaukee Area Technical College, Milwaukee, WI 54701
Moraine Park Technical Institute, Fond du Lac, WI 54935
North Central Technical Institute, Wausau, WI 54401
University of Wisconsin–Extension, Madison, WI 53706
University of Wisconsin at Oshkosh, Oshkosh, WI 54901
Waukesha County Technical Institute, Pewaukee, WI 53072
Western Wisconsin Technical Institute, La Crosse, WI 54601

Wyoming

Casper College, Casper, WY 82601
Central Wyoming College, Riverton, WY 82501
Laramie County Community College, Cheyenne, WY 82001
Northwest Community College, Powell, WY 82435
Western Wyoming Community College, Rock Springs, WY 82901

Canada

British Columbia

University of British Columbia, Centre for Continuing Education, Vancouver, British Columbia V6T 1W5. Workshops also held at Invermere, Fort Nelson, and Prince George.

Manitoba

Brandon University, Office of Continuing Education, Brandon, Manitoba R7A 6A9
University of Manitoba, Winnipeg, Manitoba R3T 3C8

New Brunswick

University of New Brunswick, Department of Extension, Box 4400, Fredericton, NB E3B 5A3

Ontario

Laurentian University, Ramsey Lake Road, Sudbury, Ontario P3E 2C6
University of Western Ontario, London, Ontario N6A 5B8

Quebec

Macdonald College, Extension Department, Box 237, Ste. Anne de Bellevue, Quebec H0A 1C0

Saskatchewan

University of Regina, Department of Extension, Regina, Saskatchewan S4S 0A2
University of Saskatchewan, Extension Division, Saskatoon, Saskatchewan S7N 0W0

Footnotes

1. Hayes, Denis, *Rays of Hope: The Transition to a Post-Petroleum World*, W. W. Norton & Co., New York, 1977, p. 139.

Part I

1. Studs Terkel, *Working*, Pantheon Press, New York, 1974.
2. Wendell Berry, *The Unsettling of America: Culture and Agriculture*, Sierra Club Books, San Francisco, 1977, p. 19.
3. Paul R. and Anne H. Ehrlich, *Population, Resources, Environment*, W. H. Freeman, San Francisco, 1972.
4. Denis Hayes is Executive Director of the Solar Energy Research Institute in Boulder, Colorado; he was formerly Chairman of Solar Lobby and a Senior Researcher at Worldwatch Institute (see p. 71).
5. Amory Lovins, *Soft Energy Paths*, Ballinger Publishing Co., Philadelphia, 1977.
6. E. F. Schumacher, *Small Is Beautiful*, Harper & Row, New York, 1973.
7. Wendell Berry, op. cit.
8. Robert Stobaugh and Daniel Yergin, *Energy Future*, Random House, New York, 1979 (Harvard Business School study). Union of Concerned Scientists, *Energy Study*, Union of Concerned Scientists, 1208 Massachusetts Avenue, Cambridge, Mass. 02138.
9. Federal Energy Administration, *Conservation Investment as a Gas Utility Supply Option*, Government Printing Office, Washington, D.C., 1976.
10. Edward J. Carlough, "A Note of Explanation," Sheet Metal Workers' International Association News Release, March 8, 1977 (see also third item in the following footnote).
11. James Benson, "Long Island Solar and Jobs Study," Council on Economic Priorities, 84 Fifth Avenue, New York, N.Y. 10011, July 1978.

    Fred Branfman and Steve LaMar, "Jobs from the Sun: Employment Development in the California Solar Energy Industry," California Public Policy Center, 304 South Broadway, Suite 224, Los Angeles, Calif., February 1978.

    H. W. Brock, G. R. Murray, J. D. McConnel, and J. C. Snipes, "Strategic Implications of Solar Energy for Employment of Sheet Metal Workers," Stanford Research Institute, Menlo Park, Calif. 94025, June 1975.

    Richard Grossman and Gail Daneker, "Jobs and Energy," Second Edition, Environmentalists for Full Employment, 1101 Vermont Avenue N.W., Washington, D.C. 20005, February 1978.

    Mid-Peninsula Conversion Project, "Creating Solar Jobs: Options for Military Workers and Communities," Mid-Peninsula Conversion Project, 967 Dana, Suite 203, Mountain View, Calif. 94041, 1979.

    New England Energy Congress, Committee on Economic Development Through Alternative Sources of Energy, "Jobs and Energy," edited transcripts of public hearings held before the U.S. Senate Subcommittee on Jobs, Fall 1978, New England Energy Congress, 14 Whitfield Road, Somerville, Mass. 02144.

    Leonard S. Rodberg, "Employment Impact of the Solar Transition," Report of Subcommittee on Energy, Joint Economic Committee of the U.S. Congress, April 1979.

    Meg Schachter, "The Job Creation Potential of Solar and Conservation: A Critical Evaluation," U.S. Department of Energy, Advanced Energy Systems Division, Forrestal Building, Room 6E068, 1000 Independence Avenue S.W., Washington, D.C. 20009, November 1978.

Lee Webb, "Full Employment through a National Solar Energy/Energy Conservation Program: First Steps Toward a Legislative Proposal," Conference on Alternative State and Local Policies, 1901 Q Street N.W., Washington, D.C. 20009, 1978.

Thomas A. Welch, Robert J. Perello, and Jan. M. Dyroff, "Occupational Impact of Solar Energy," Massachusetts Occupational Information System, 60 Williams Street, Wellesley Hills, Mass. 02181, September 1977.

12. *Worklife*, U.S. Department of Labor, August 1976.
13. Jerry Flint, "Oversupply of Young Workers Expected to Tighten Jobs Race," *New York Times*, June 24, 1978, p. 1.

Part II

1. Roy Reed, "Rural Areas' Population Gains Now Outpace Urban Regions," *New York Times*, May 18, 1975, p. 1.
2. Bruce Anderson, with Michael Riordan, *The Solar Home Book*, Brick House Publishing Company, Harrisville, N.H., 1976.
3. E. F. Schumacher, op. cit.
4. "Turning an Ideology into a Practical Solution," *Maine Times*, January 14, 1976.

Part III

1. "The Coming Boom in Solar Energy," *Business Week*, October 9, 1978, p. 102.
2. American Gas Association Solar Energy Committee, *Solar Energy Utilization*, American Gas Association, 1515 Wilson Boulevard, Arlington, Virginia 22209, November 1977.
3. Dun and Bradstreet, *Million Dollar Directory*, Vol. I, Dun and Bradstreet, Inc., 3 Century Drive, Parsippany, N.J. 07054, 1979.
4. Ford Foundation, *A Time to Choose*, Ballinger Publishing Company, Cambridge, Mass., 1974.
5. Citizens Energy Project, *Industrial Energy Conservation*, Report Series No. 25, Citizens Energy Project, 1413 K Street NW, 8th Floor, Washington, D.C. 20005, 1978.
6. From International Association of Machinists & Aerospace Workers; Sheet Metal Workers; International Association, and Oil, Chemical and Atomic Workers (see also #8 below).
7. See Part I, #11.
8. *Joint Economic Committee Holds Jobs and Energy Hearings* (report), Environmentalists for Full Employment, 1101 Vermont Avenue N.W., #305, Washington, D.C. 20005, Spring 1978.
9. E. F. Schumacher, op. cit.
10. Ken Bossong, *Solar Cells*, Citizens Energy Project, 1413 K Street N.W., 8th Floor, Washington, D.C. 20005.
11. *The First Year*, annual report, Solar Energy Research Institute, 1536 Cole Boulevard, Golden, Colorado 80401, December 1978, p. 2
12. *Solar Age*, official magazine of the American section of the International Solar Energy Society, Inc. Subscriptions: P.O. Box 4934, Manchester, N.H. 03108.

    Solar Engineering Magazine, official publication of the Solar Energy Industries Association. Subscriptions: 8435 N. Stemmons Freeway, Suite 880, Dallas, Texas 75247.
13. Solar energy curriculum for grades 7 through 12: Solar Energy Education Project, Bureau of Science Education, The State Education Department, Room 302 EB, The University of the State of New York, Albany, N.Y. 12234.

    Solar energy curriculum for elementary schools: Dr. Seymour Lampert, Department of Mechanical Engineering, University of Southern California, University Park, Los Angeles, Calif. 90007.

14. Bill Trussell (editor), with Dr. Maria Dalton and Susan Conway (principal investigators), *The Energy Education Materials Inventory, Vol. II*, prepared at the University of Houston (Texas) Energy Laboratory, with funding by the U.S. Department of Energy. Available from Superintendent of Documents, U.S. Government Printing Office, Washington, D.C. 20402.

    Minnesota Energy Agency, *Energy Education Materials Bibliography*, available from ERIC, Ohio State University, 1200 Chambers Street, 3rd Floor, Columbus, Ohio 43212, 1978 (being updated). Prepared in response to requests from teachers, this bibliography reviews the 100 materials the agency considered best, evaluating them for content, subject areas, grade level, points of emphasis, and teacher background necessary; identifies the approach and any apparent biases.
15. Arnold R. Wallenstein, *Barriers and Incentives to Solar Energy Development: An Analysis of Legal and Institutional Issues in the Northeast*, Northeast Solar Energy Center, 70 Memorial Drive, Cambridge, Mass. 02142, December 1978 (with an appendix analyzing the solar and energy conservation provisions of the 1978 National Energy Act).
16. Mitre Corp., "Solar Energy Technology Delivery Systems: Water and Space Heating for Residential and Commercial Buildings," *Solar Energy Technology Transfer Source Book*, Vol. I (Technical Report MTR-7674), Metrek Division of the Mitre Corp., 1820 Dolley Madison Boulevard, McLean, Virginia 22102, August 1978.
17. *The Solar Law Reporter*, published by Solar Energy Research Institute (SERI). Subscriptions: P.O. Box 5400, Denver, Colorado 80217.
18. William A. Thomas, Alan S. Miller, and Richard L. Robbins, *Overcoming Legal Uncertainties About Use of Solar Energy Systems*, American Bar Foundation, 1155 East 60th Street, Chicago, Illinois 60637, 1978 (see also #15, above).

Part IV

Research sources for Part IV include the *directories* listed, publications put out by many of the small solar-related *organizations*, and information gathered from the *government special solar programs*.

The author also obtained or validated listings by direct contact, often through correspondence or personal interviews with personnel in programs mentioned. In addition, each solar organization checked the copy to be used; each federal office and agency was reached by telephone to assure the accuracy of the information; and each college and university supplied catalogues and special publications.